The Elements of Mathematics from a Modern Viewpoint II
The Problems and their Solutions

Jean-Claude Falmagne

Elementary Number Theory
Rational Numbers
Set Theory
Real Numbers
Elementary Algebra
Geometry
Probability Theory
Statistics

Preamble

This is a companion volume of *The Elements of Mathematics I*. It contains almost all of the solutions of the problems proposed in Volume I. For convenience, the statements of the problems are recalled here, and so are some definitions and formulas required to understand the solutions. As in Volume I, the problems are presented in a yellow frame, followed by the solutions. Before each frame, the section number and the page where the problems appeared in Volume 1, are indicated. An example is given below.

Section 2.4.

Page 35.

> **Problems**
>
> In each case, keeping your eyes closed, multiply:
>
> 1. 91×99;   2. 52×58;   3. 77×73;   4. 85×85;   5. 111×119.
>
> (This would be a good time to learn your multiplication table up to 12.)

SOLUTIONS

1. $9,009$.   2. $3,016$.   3. $5,621$.   4. $7,225$.   5. $13,209$.

Also as in Volume I, the problems intended for the "**very curious students**" appear in a green frame in a wiggly surround. Example:

> **Problems for the very curious students.**

In the Appendix, on pages 143-146, the abbreviations of the units of length, mass, capacity, and time are recalled. This volume is practically self contained. It can be used without reference to Volume I by a student knowing the subject and wanting to assess his or her competence in the K-12 mathematics curriculum.

I am much grateful to Ava Soleimany, Hussain Jafari, and Kavi Sakraney, for their important contributions to the solutions of the problems. They also provided some of the graphics in this volume.

JCF

Table of Content

Chapter 1.	Whole Numbers—Ordering and Adding..	page 1
Chapter 2.	Multiplying, Dividing and Counting......	page 11
Chapter 3.	The Integers.............................	page 31
Chapter 4.	The Rational Numbers....................	page 45
Chapter 5.	Sets, Relations and Functions	page 55
Chapter 6.	Real Numbers	page 66
Chapter 7.	Linear Equations and Linear Inequalities	page 75
Chapter 8.	Non Linear Equations	page 87
Chapter 9	Lines and Angles.........................	page 100
Chapter 10.	Triangles and Quadrangles	page 108
Chapter 11.	Polygons and the Circle	page 113
Chapter 12.	Three-Dimensional Figures	page 119
Chapter 13.	Probability Theory	page 122
Chapter 14.	Statistics...............................	page 134
Appendix.	....	page 143

Chapter 1

Whole Numbers—Ordering and Adding

Section 1.2.

Page 3.

> **Problems**
>
> In the whole number grid below, the thick red point marks the ordered pair $(4,6)$. We do this because this point is the crossing point of the vertical whole number line marked 4 and the horizontal whole number line marked 6.
>
> 1. On this whole number grid, mark by a big black dot all the points corresponding to the following ordered pairs:
>
> $(1,7)$, $(7,1)$, $(7,3)$, $(4,4)$, $(6,5)$, and $(5,6)$.
>
> 2. On the whole number grid, what is special about all the points (m, n) such that
>
> $m - n = 1$?
>
>   One of them is $(3, 2)$.

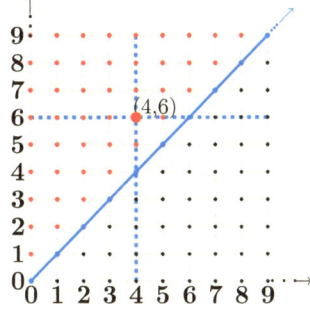

SOLUTIONS

1. The ordered pairs $(1,7)$, $(7,1)$, $(7,3)$, $(4,4)$, $(6,5)$, and $(5,6)$ are marked by thick black dots on the grid below on the right.

2. The ordered pairs (m, n) such that $m - n = 1$ lie on the black line. The ordered pair $(6, 5)$ is contained in that line. The blue line contains all the ordered pairs (m, n) such that $m - n = 0$.

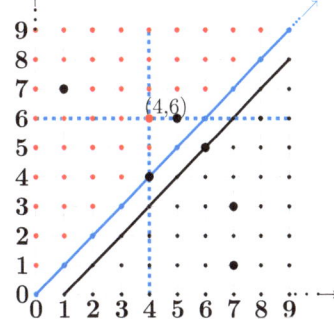

1

Page 5.

Problems

In the whole number grid below, two thick black points have been marked to indicate the two ordered pairs $(2,3)$ and $(5,9)$. Draw the line joining these two points and extending to the end of the grid.

1. This line meets some other points of the grid. List the ordered pairs corresponding to these points.

2. **For the very curious students.** Examine all the ordered pairs from Problem 1. Can you think of a rule that would generate all the ordered pairs (m,n) marking the points of the line? **Hint:** This rule has the form $k \times m - j = n$, where k and j are whole numbers.

3. Suppose that you extend the grid and the line upward and to the right. You are told that the ordered pair $(20, n)$ corresponds to a point of the line. What would then be the value of n?

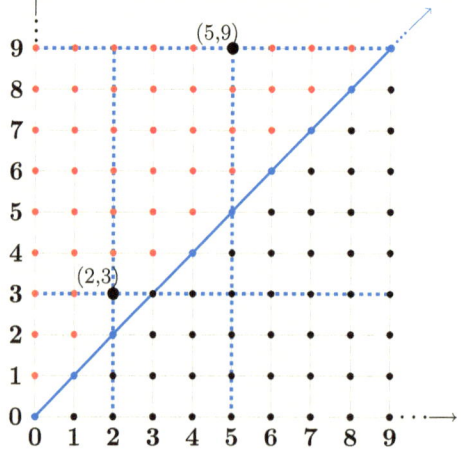

Solutions

1. $(1,1)$, $(3,5)$, $(4,7)$.

2. The rule is: $2m - 1 = n$.

3. 39. The point is thus $(20, 39)$.

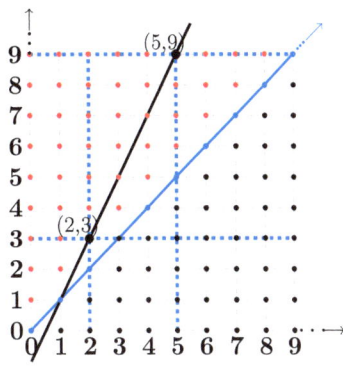

Page 7.

Problems

1. Suppose that, for a whole number n, you know that $7 < n$ and $n < 11$. What then are the possibilities for the number n?

2. Let m and 3 be two whole numbers. What you know about m is that neither $3 < m$ nor $m < 3$ is true. What is m? Which law, or property, about the whole numbers are you using to find out?

3. The transitivity property for the 'greater than' relation $>$ says that, for any whole numbers m, n and k, the following is true: if $m > n$ and $n > k$, then $m > k$. Give three examples of that property for numbers m, n and k smaller than 10.

4. The figure below, on the right, gives a picture of the transitivity property. We have $1 < 4$ and $4 < 8$, which gives $1 < 8$. The three red points $(1,4)$, $(4,8)$ and $(1,8)$ correspond to $1 < 4$, $4 < 8$, and $1 < 8$, forming the red triangle.

The last point $(1,8)$ is the crossing point of the vertical side and the horizontal side of the triangle. Construct the same kind of triangle for the three cases:

- if $5 < 6$ and $6 < 9$, then $5 < 9$;
- if $2 < 5$ and $5 < 8$, then $2 < 8$;
- if $1 < 6$ and $6 < 7$, then $1 < 7$.

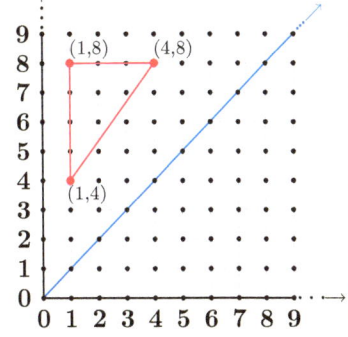

SOLUTIONS

1. 8, 9 and 10.

2. $m = 3$ by the Law of Trichotomy.

3. By transitivity:

   $2 < 3$ and $3 < 5$ implies $2 < 5$;

   $1 < 4$ and $4 < 6$ implies $1 < 6$;

   $4 < 6$ and $6 < 9$ implies $4 < 9$.

4. The transitivity triangles for the three examples.

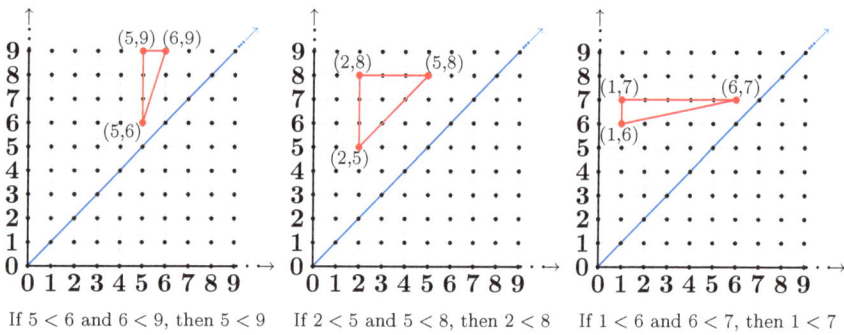

If $5 < 6$ and $6 < 9$, then $5 < 9$ If $2 < 5$ and $5 < 8$, then $2 < 8$ If $1 < 6$ and $6 < 7$, then $1 < 7$

Page 8.

> **Three problems for the very curious students.**
>
> In the three problems below, we write $m \ll n$ to mean two things:
> - $m < n$;
> - and there is at least one whole number between them; that is, there is some number k such that $m < k$ and $k < n$.
>
> For example, we have $3 \ll 6$ because $3 < 4$ and $4 < 6$. We could call the relation $\ll$ *'quite smaller than'*.
>
> 1. Replace five of the points in the grid on the next page by the symbol $\ll$. Thus, in each case, replace the black dot of the point (m, n) by $\ll$, which means: $m \ll n$.
>   (Hint: all the points to be replaced by $\ll$ are above the diagonal; one point has been replaced already.)

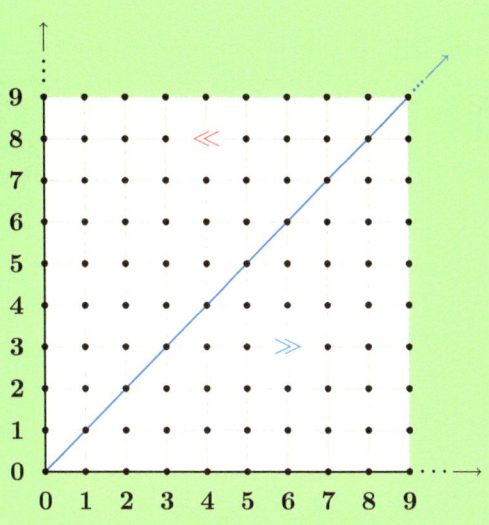

2. Continuing with Problem 1: Replace five points of the grid by the relation $\gg$, which means *much greater than*. Thus, we have $m \gg n$ if $m > n$ and there is at least one number between them.

3. Continuing with Problem 1: How would you describe the part of the grid with all the points that could not be marked by either $\ll$ or $\gg$? This is also a relation. What would be a good name for it? Are the relations $\ll$ and $\gg$ also transitive? Look at a few cases and make a guess. Try to come up with a proof that your guess is correct.

SOLUTIONS

1 and 2.

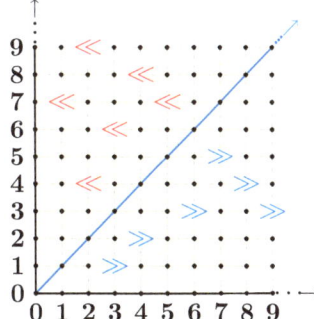

6   CHAPTER 1. WHOLE NUMBERS—ORDERING AND ADDING

3. **Continuation.** The points that could not be marked by either $\gg$ or $\ll$ are either on the blue line, or next to it. All the ordered pairs marking points on the blue line are of the form (m,m). The other points next to the blue line are marked by either $(m,m+1)$ or $(m,m-1)$. So, we could say that these ordered pairs (m,n) are such that with m *differs from n by at most* 1. The two relations $\gg$ and $\ll$ are transitive. In particular, the transitivity of $\ll$ results from the transitivity of $<$. For example, $2 \ll 5$ and $5 \ll 9$ imply $2 \ll 9$: we have $2 < 3 < 5$ and $2 < 3 < 9$, and so $2 \ll 9$. The proof of the transitivity of $\ll$ uses the same idea. Suppose that $m \ll n$ and $n \ll q$. Then $m < k < n < r < q$ for some whole numbers k and r. So $m < k < q$, which gives $m \ll q$. The proof of the transitivity of $\gg$ is similar.

Section 1.3.

Page 12.

> **Problems**
>
> 1. How many ordered pairs (j,k) of whole numbers are there with a sum $j+k$ equal to 8? (Hint: you can simply count the red 8's in Figure 1.1.)

> **A problem for the very curious students.**
>
> 2. For a whole number n, how many ordered pairs (k,m) of whole numbers are there such that $k+m=n$? Can you find a general rule for computing such a number? (Hint: look at the two cases $n=4$ and $n=5$, and try to generalize.)

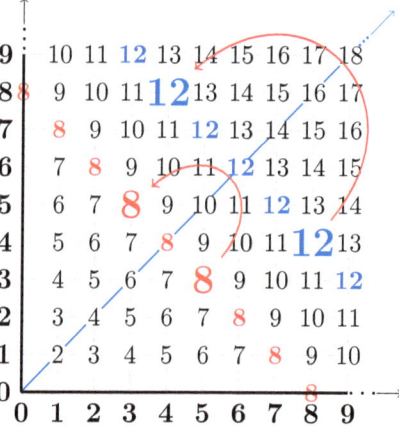

Figure 1.1: The addition grid. The two curved red arrows mark the effect of folding along the blue diagonal, picturing the two commutativity equations:

$4+8=8+4$ and $3+5=5+3$.

To simplify the picture, we have omitted adding with 0, except for the two ordered pairs $(0,8)$ and $(8,0)$, giving $8+0=0+8=8$.

SOLUTIONS

1. There are $8+1=9$ such pairs: four above the blue line, four below it, and the pair $(4,4)$ on the line.

2. For the very curious students. The rule is that there are $n+1$ ordered pairs (k,m) such that $m+k=n$. This is true whether m is even or odd. Check with the cases $n=4$ and $n=5$.

Section 1.4.

Page 15.

Problems

1. Compute:

(a) 488
    8576
  +   24

(b) 7499
      498
  +   65

(c) 1825
       73
  + 282

(d) 2737
       79
  + 258

(e) 7399
      936
  +   87

(f)   663
     6511
  +   78

(g) 3717
      758
  +   75

(h) 2619
      448
  +   36

Solve the two word problems:

2. Yesterday, Bill had 472 baseball cards. Today, he got 185 more. How many cards does Bill have now?

3. Karen has 604 beads and Bob has 218 beads. How many beads do Karen and Bob have together?

SOLUTIONS

1.
   (a) $9,088$; (b) $8,062$; (c) $2,180$; (d) $3,074$;
   (e) $8,422$; (f) $7,252$; (g) $4,550$; (h) $3,103$.

2. 657.

3. 822

Section 1.6.

Page 19.

> **Problems**
>
> 1. Compute:
>
> (a) 82 (b) 68 (c) 76 (d) 87
> -43 -29 -68 -79
>
> (e) 70 (f) 55 (g) 41 (h) 92
> -39 -36 -19 -29
>
> Solve the word problems:
>
> 2. Last year, a bake sale made \$321. This year, the bake sale made \$947. How much more money did the bake sale make this year?
>
> 3. Karen has 604 beads and Bob has 218 beads. How many more beads does Karen have?

SOLUTIONS

1. (a) 39; (b) 39; (c) 8; (d) 8;
   (e) 31; (f) 19; (g) 22; (h) 63.
2. 626.   3. 386.

Page 21.

> **Problems**
>
> Compute the following subtractions.
>
> (a) 435 (b) 111 (c) 567 (d) 304
> -86 -23 -78 -286
>
> (e) 534 (f) 200 (g) 778 (h) 453
> -67 -89 -79 -286
>
> (i) 5003 (j) 1616 (k) 9044
> -689 -888 -268

SOLUTIONS

(a) 349;   (b) 88;   (c) 489;   (d) 18;   (e) 467;   (f) 111;

(g) 699;   (h) 167;   (i) 4,314;   (j) 728;   (k) 8,776.

Section 1.8.

Page 23, 24.

> **Review Problems**
>
> 1. If $3 \leq m$ and $m \leq 9$, what are the possible values for m?
>
> 2. If $(m, 7)$ is a point above the diagonal on the whole number grid, what are the possible values for m?
>
> 3. The points $(2, 5)$, $(5, 6)$, and $(8, 1)$ have been marked by red dots in the whole number grid displayed in Problem 4. Mark the same way by ordered pairs and red dots the points $(7, 7)$, $(1, 7)$, and $(6, 0)$.
>
> 4. In one of the figures of this chapter, we have replaced each of the points of the grid by one of $<, =,$ and $>$. Without searching for that figure, in the whole number grid on the right, replace each of the five blue points by one of $<, =,$ and $>$.
>
> 5. What are the ordered pairs (m, n) of whole numbers m and n such that: $2 \leq m \leq n \leq 5$?
>
> 6. If $m - k < 8 - k$, what are the possible values for the whole numbers k and m?

SOLUTIONS

1. 3, 4, 5, 6, 7, 8, and 9.

2. 0, 1, 2, 3, 4, 5, and 6.

3 and 4. See the figure on the right.

5. $(2, 2), (2, 3), (2, 4), (2, 5),$
   $(3, 3), (3, 4), (3, 5),$
   $(4, 4), (4, 5),$
   $(5, 5)$.

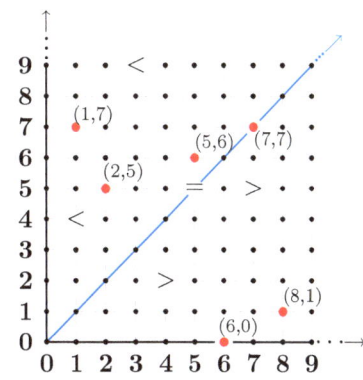

6. We must have $k < 8$ because if $k = 8$, then $m - 8 < 0$, and so $m - 8$ cannot be an whole number. We cannot have $m = 8$, because this leads to $8 - k < 8 - k$. So, either $k < m$ or $m < k$. In fact, we must have $m < 8$, since $m - k < 8 - k$. To sum up, we have $k \leq m$, $k < 8$ and $m < 8$. The list of all the ordered pairs (m, k) of whole numbers satisfying these three conditions is:

$(7, 7), (7, 6), (7, 5), \ldots, (7, 0),$

$(6, 6), (6, 5), (6, 4), \ldots, (6, 0),$

$(5, 5), (5, 4), (5, 3), \ldots, (5, 0),$

$(4, 4), (4, 3), (4, 2), \ldots, (4, 0),$

$(3, 3), (3, 2), (3, 1), (3, 0),$

$(2, 2), (2, 1), (2, 0),$

$(1, 1), (1, 0),$

$(0, 0).$

Chapter 2

Whole Numbers—Multiplying, Dividing, and Counting

Section 2.1.

Page 28.

> **Problems**
>
> 1. Imagine that we build a multiplication grid in the style of the addition grid of Figure 1.1, but much bigger, for example a 25×25 grid. How many times will the number 25 appear in the grid?
>
> 2. No matter how large we make the multiplication grid, there will be some numbers that will appear only in two points. The number 15 is an example: it appears only at the four points $(3,5)$, $(5,3)$, $(1,15)$, and $(15,1)$. On the other hand, the number 12 appears 7 times. What are some other examples of numbers that appear only twice in the table? What is special about these numbers?

SOLUTIONS

1. Three times, corresponding to the three products: 1×25, 25×1, and 5×5.

2. What is special about these numbers is that the only kind of products producing them are of the form $1 \times n$ and $n \times 1$. Some examples are $1 \times 7 = 7 \times 1 = 7$ and $1 \times 17 = 17 \times 1 = 17$.

Page 29.

> **Problems**
> 1. Remove the unnecessary parentheses in the expressions below.
>   (a) $((145 \times 11) - (6 \times 2)) + 2 \times (3 \times 4)$.
>   (b) $13 \times (4 + 2) - (3 \times 6)$.
> 2. Add the necessary parentheses to make:
>   (a) $7 \times 8 - 4 \times 3 + 5$ equals 224;   (b) $9 - 7 - 5 \times 2 + 1$ equals 21.
> 3. Compute $(8 - 6) \times (2 + 5 \times 8)$.

SOLUTIONS

1. (a) $145 \times 11 - 6 \times 2 + 2 \times 3 \times 4$;   (b) $13 \times (4 + 2) - 3 \times 6$.
2. (a) $7 \times ((8 - 4) \times (3 + 5)) = 224$;   (b) $(9 - (7 - 5)) \times (2 + 1) = 21$.
3. $(8 - 6) \times (2 + 5 \times 8) = 84$.

Section 2.2.

Page 32.

> **Problems**
> Solve the word problems:
> 1. There are 4 piles of papers on a desk. Each pile has 12 papers. How many papers are there in total?
> 2. A group of 16 friends went to a sporting event. It cost \$32 for each of them to get in. What was the total cost for all of them to get in?
>
> Compute the multiplications below:
>
> 3. $\begin{array}{r} 114 \\ \times\ \ \ \ 5 \\ \hline \end{array}$ 4. $\begin{array}{r} 57 \\ \times\ \ 8 \\ \hline \end{array}$ 5. $\begin{array}{r} 66 \\ \times\ \ 5 \\ \hline \end{array}$
>
> 6. $\begin{array}{r} 33 \\ \times\ 31 \\ \hline \end{array}$ 7. $\begin{array}{r} 31 \\ \times\ 13 \\ \hline \end{array}$ 8. $\begin{array}{r} 42 \\ \times\ 12 \\ \hline \end{array}$
>
> 9. $\begin{array}{r} 38 \\ \times\ 48 \\ \hline \end{array}$ 10. $\begin{array}{r} 649 \\ \times\ \ 79 \\ \hline \end{array}$ 11. $\begin{array}{r} 866 \\ \times\ \ 98 \\ \hline \end{array}$

SOLUTIONS

1. 48 papers. 2. $512.
3. 570. 4. 456. 5. 330.
6. 1,023. 7. 403. 8. 504.
9. 1,824. 10. 51,271. 11. 84,868.

Section 2.3.

Page 33.

Problems

Rewrite each expression using the distributive property. Then, evaluate the expression:

1. $3 \times (4 + 7)$;
2. $4 \times (5 + 8)$;
3. $(2 + 3) \times 7$;
4. $37 \times (100 + 10)$.

Transform each expression into a single product by distributivity.

5. $3 \times 3 + 3 \times 4 + 3 \times 6$;
6. $7 \times 2 + 7 \times 4 + 7 \times 6$;
7. $5 \times 5 + 5 \times 5 + 5 \times 5 + 5 \times 5$.

SOLUTIONS

1. $3 \times 4 + 3 \times 7 = 33$.
2. $4 \times 5 + 4 \times 8 = 52$.
3. $2 \times 7 + 3 \times 7 = 35$.
4. $37 \times 100 + 37 \times 10 = 4{,}070$.
5. $3 \times (3 + 4 + 6) = 3 \times 13 = 39$.
6. $7 \times (2 + 4 + 6) = 7 \times 12 = 84$.
7. $5 \times (5 + 5 + 5 + 5) = 5 \times 20 = 100$.

Section 2.4.

Page 35.

> **Problems**
>
> In each case, keeping your eyes closed, multiply:
>
> 1. 91×99;   2. 52×58;   3. 77×73;   4. 85×85;   5. 111×119.
>
> (This would be a good time to learn your multiplication table up to 12.)

SOLUTIONS

1. $9,009$.   2. $3,016$.   3. $5,621$.   4. $7,225$.   5. $13,209$.

Page 36.

> **Problems**
>
> Multiply:
>
> 1. 91×11;   2. 46×11;   3. 83×11;   4. 57×11;   5. 38×11.

SOLUTIONS

1. $1,001$.   2. 506.   3. 913.   4. 627.   5. 418.

Section 2.5.

Page 39.

> **Problems**
>
> Remove all the parentheses that are not needed in the three expressions below. In each case, carry out the calculations.
>
> 1. $\big((6 \div 2) \times (5 \times 8) + (18 \div 3)\big)$;
> 2. $\big(15 \div (2 + 1)\big) - (9 \div 3)$;
> 3. $\big((4 + 2) \div (3 - 1)\big) + 6$.
>
> In each of the following three cases, add the necessary parentheses so that the result indicated becomes correct.
>
> 4. $1 + 11 \div 2 + 2 = 3$;
> 5. $6 \times 2 + 1 \div 3 \div 2 = 3$;
> 6. $5 \times 21 \div 3 + 4 + 81 \div 4 + 5 = 48$. (This one may take some time.)

SOLUTIONS

1. $6 \div 2 \times 5 \times 8 + 18 \div 3 = 126.$
2. $15 \div (2+1) - 9 \div 3 = 2.$
3. $(4+2) \div (3-1) + 6 = 9.$
4. $(1+11) \div (2+2) = 3.$
5. $\big(6 \times (2+1) \div 3\big) \div 2 = 3.$
6. $5 \times 21 \div 3 + 4 + 81 \div (4+5) = 48.$

Section 2.6.

Page 41.

> **A problem for the very curious students.**
>
> Find the smallest whole number m which gives a remainder of 2 when divided by three different whole numbers.

SOLUTION

The smallest number is 14. The numbers dividing 14 are 2, 3 and 6.

> **Problems**
>
> Compute the following 8 divisions, indicating the remainder, if there is one. For example, the answer to Problem 4 is: $307 = 4 \times 76 + 3$. So 3 is the remainder.
>
> 1. $718 \div 2$ 2. $98 \div 7$ 3. $876 \div 6$ 4. $307 \div 4$
> 5. $483 \div 13$ 6. $993 \div 22$ 7. $19653 \div 16$ 8. $8234 \div 212$.
>
> 9. A pet store has 12 tanks of fish, and each tank has the same number of fish. There is a total of 180 fish at the pet store. How many fish are in each tank?
>
> 10. Chang has to carry 325 apples from a farm to the market. How many baskets will he need, if each basket can hold 41 apples?

SOLUTIONS

1. 359.
2. 14.
3. 146.
4. $307 = 4 \times 76 + 3.$
5. $483 = 13 \times 37 + 2.$
6. $993 = 22 \times 45 + 3$
7. $19653 = 16 \times 1228 + 5.$
8. $8234 = 212 \times 38 + 178$
9. 15 fish.
10. Chang needs 8 baskets. Seven of them with 41 apples, and the last one 38 apples. Here 38 is the remainder of the division of 325 by 7.

Sections 2.7 and 2.8.

Page 44.

> **Problems**
> Evaluate the following expressions:
> 1. $9 + 5 \times 9$
> 2. $15 \div 5 - 2$
> 3. $48 - 24 \div 6$
> 4. $56 + 3 \times 4$
> 5. $8 + 4 \times 6 + 9 \div 3$
> 6. $14 \div 2 - 5 + 4 \times 7$
> 7. $4 \times 6 - 8 - 8 \div 2$
> 8. $5 + 4 \times 7 - 12 \div 3$
> 9. $(914 - 120 \times 5 + 14) \div 2$
> 10. $6 \times ((14 + 1) \div 3 + 2)$
> 11. $5 \times (5 + (15 + 17) \div 8)$
> 12. $(17 + 5 \times (13 - 10)) \div 4$

SOLUTIONS

1. 54.   2. 1.   3. 44.   4. 68.   5. 35.   6. 30.
7. 12.   8. 29.   9. 164.   10. 42.   11. 45.   12. 8.

Section 2.9.

Pages 44 and 45.

> **Problems**
> 1. Find all the divisors of each the following numbers:
>   (a) 27;   (b) 54;   (c) 64;   (d) 11;   (e) 31.
> 2. What is special about the last two, 11 and 31? Can you think about other numbers like that? Is 64 also special?
> 3. Factor each of the following numbers in all possible ways:
>   (a) 6;   (b) 33;   (c) 57.
> 4. What are all the common divisors of 24, 36 and 64?
> 5. Find two numbers that have no common divisors except 1, but each of them has at least 3 divisors other than 1.

SOLUTIONS

1. The divisors of each of the numbers are:
    (a) $1, 3, 9$ and 27; (b) $1, 2, 3, 6, 9, 18, 27$ and 54;
    (c) $1, 2, 4, 8, 16, 32$ and 64; (d) 1 and 11; (e) 1 and 31.

2. The numbers 11 and 31 are special: they only have 1 and themselves as divisors. The numbers 3 and 5 are also special in the same way. That is not the case for 64, whose divisors other that 1 and 64 are $2, 4, 8, 16$ and 32. But 64 is special in that its divisors are all multiples of 2.

3. (a) $6 = 1 \times 6 = 6 \times 1$
    $= 1 \times 2 \times 3 = 1 \times 3 \times 2$
    $= 2 \times 1 \times 3 = 2 \times 3 \times 1$
    $= 3 \times 1 \times 2 = 3 \times 2 \times 1$
   (b) $33 = 1 \times 33 = 33 \times 1;$
    $= 3 \times 11 = 11 \times 3;$
   (c) $57 = 1 \times 57 = 57 \times 1.$

4. The common divisors of 24, 36 and 64 are $1, 2$ and 4.

5. Two such numbers are $78 = 2 \times 3 \times 13$ and $385 = 5 \times 7 \times 11$.

Section 2.10.

Page 50.

Problems

Using the Laws of Exponents, evaluate the following expressions.

1. 2^5. 2. 3^4. 3. 13^2. 4. 5^3.
5. $\frac{4^3}{2^2}$. 6. 5^{4-3}. 7. $\frac{12}{2^2}$. 8. $\frac{7^5}{7^3}$.
9. $\frac{4^2}{2}$. 10. $\frac{2^4}{4^2}$. 11. $\frac{16^4}{8^5}$. 12. $\frac{6^3}{3^2}$.

Using the Laws of Exponents, simplify the following expressions as much as possible.

13. $w^2 \times w^4 \times w$. 14. $\frac{u^5}{u^4}$.
15. $\frac{3ab^7}{a^6 b^4}$. 16. $\frac{y^2 z^5}{y^2 z^4 b^3}$.
17. $4w^7 \cdot 2v \cdot 2v^9 w^4$. 18. $4x^5 \cdot 3u^7 x^5 \cdot 2u$.
19. $6u^5 \cdot 4y^5 u^3 \cdot 2y$. 20. $2u^7 \cdot 8w^9 u^3 \cdot 2w$.
21. $(2yz^4)^4$. 22. $\left(\frac{3a^4}{b^3}\right)^5$.
23. $(v^7)^6$. 24. $(3z^4 x)^3$.

Using the Laws of Exponents, first simplify, and then compute the expressions below.

25. $\frac{18^{5-2}}{3^{6-4}}$. 26. $\frac{15^{6-3}}{3^{7-4}}$. 27. $\frac{8^{2+1}}{4^{6-3}}$. 28. $\frac{(3^2)^6}{3^{11}}$.

29. $\frac{6^4 \times 6^5}{2^9 \times 3^9}$. 30. $\frac{6^6 \times 3^6}{2^5 \times 9^5}$. 31. $\frac{7^3}{4^2} \times \frac{4^3}{7^2}$. 32. $\frac{16^3}{(2^3)^4}$.

SOLUTIONS

1. 32. 2. 81. 3. 169. 4. 125.
5. 16. 6. 5. 7. 3. 8. 49.
9. 8. 10. 1. 11. 2. 12. 24.
13. w^7. 14. u. 15. $\frac{3b^3}{a^5}$. 16. $\frac{z}{b^3}$.
17. $16w^{11}v^{10}$. 18. $24x^{10}u^8$. 19. $48u^8y^6$. 20. $32u^{10} \cdot 10w^{10}$.
21. $16y^4z^{16}$. 22. $\frac{243a^{20}}{b^{15}}$. 23. v^{42}. 24. $27z^{12}x^3$.

We give below the details of the solution to Problem 25.

$$\frac{18^{5-2}}{3^{6-4}} = \frac{18^3}{3^2} = \frac{18 \times 18^2}{9} = \frac{2 \times 9 \times 18^2}{9} = \frac{2 \times \not{9} \times 18^2}{\not{9}} = 2 \times 18^2 = 648,$$

with the last step obtained from the calculator or by pencil calculation.

25. 648. 26. 125. 27. 8. 28. 3.
29. 1. 30. 18. 31. 28. 32. 1.

Section 2.11.

Page 54.

We recall that the first twenty prime numbers are:

2, 3, 5, 7, 11, 13, 17, 19, 23, 29, 31, 37, 41, 43, 47, 53, 59, 61, 67, 71.

Problems

In Problems 1 to 6, give the prime factorization of each number. In each case, you must also construct the corresponding tree. (You can draw short trees if you want). You may use the list given above as a help to find the largest divisor to be checked.

1. 99. 2. 28. 3. 136. 4. 51. 5. 159. 6. 329.

7. If you have access to the internet, go to the website

    http://www.rsok.com/~jrm/next_prime.html

    to find the next prime number after each of the following numbers: 766, 336 and 1321. This program will actually compute it for you, and it may take some time.
    You can also go to the website http://primes.utm.edu, which will give you the list of the first 1000 prime numbers.

8. Using the trial division, check that each of the three next prime numbers that you have obtained in solving Problem 7 is in fact prime number.

9. If m is a prime number, can $n = m^2$ have any prime factor (or factors) other than m?

Continuation of Problem 9. Try to prove your response. Hint: you can use the fact that if j and k are prime factors of some whole number n, then j is a prime factor of $\frac{n}{k}$.

SOLUTIONS

1. $99 = 3 \times 3 \times 11$.

2. $28 = 2 \times 2 \times 7$.

```
99 ← Try 2 fails
↙ ↘
2    99 ← Try 3
     ↙ ↘
     3   33 ← Try 3 again
         ↙ ↘
         3   11 ← Try 3 fails
             ↙ ↘
             3   11 ← Try 5, 7 fail
                 ↙ ↘
                 5 7  11 ← Try 11
                     ↙ ↘
                    11   1
```

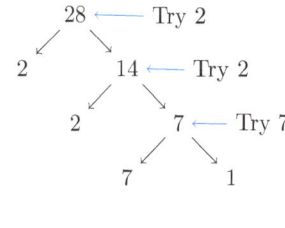

20   CHAPTER 2. MULTIPLYING, DIVIDING AND COUNTING

3. $136 = 2 \times 2 \times 2 \times 17$.

4. $51 = 3 \times 17$.

In Problem 3 and 4, after getting 17, we could have tried 3, and then 5 and 7, but there was no need because we know that 17 is a prime number. It would be good for you to know by heart the first 20 prime numbers.

5. $159 = 3 \times 53$.

6. $329 = 7 \times 47$.

7. The first prime number after 766 is 769, and the first prime number after 336 is 337. The number 1321 is itself a prime number. The next one is 1327.

8. Note that we have

$$(23)^2 = 569 < 769 < 841 = (29)^2,$$
$$(17)^2 = 289 < 337 < 361 = (19)^2,$$
$$(31)^2 = 961 < 1327 < 1369 = (37)^2,$$

with $(23, 29)$, $(17, 19)$, and $(31, 37)$ being ordered pairs of consecutive prime numbers. The first twelve prime numbers are

2, 3, 5, 7, 11, 13, 17, 19, 23, 29, 31, 37.

We give on the next page the three short division trees verifying that 769, 337 and 1321 are prime numbers:

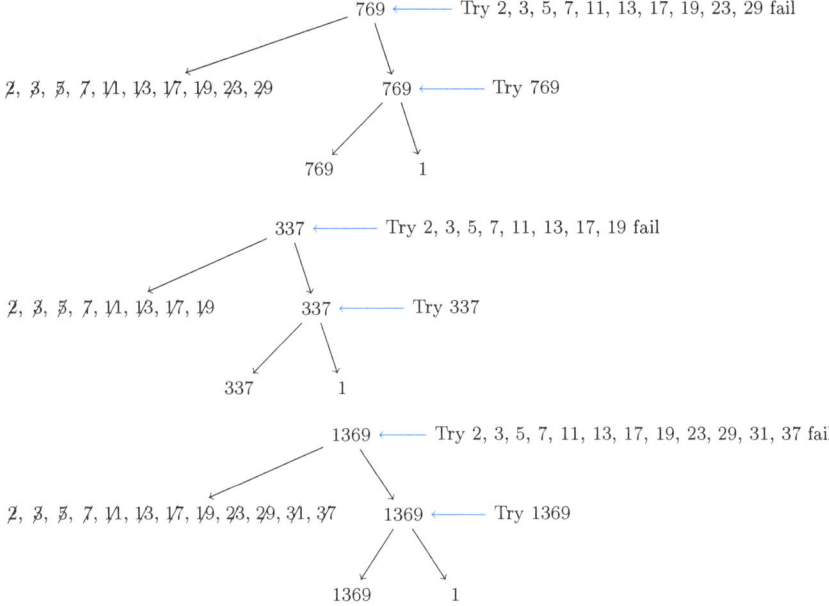

9. If m is a prime number, then the only prime factor of $n = m^2$ is m. For the very curious students, we first prove: *If j and k are prime factors of n, then j is a prime factor of $\frac{n}{k}$.*

This is true because if j and k are prime factors of n, then $\frac{n}{jk} = p$ for some whole number p, and so j is a prime factor of $\frac{n}{k} = jp$. Now suppose that j is a prime factor of $n = m^2$ other than m. Then using the hint, j must be a prime factor of $\frac{n}{m} = \frac{m^2}{m} = m$ different from m. This cannot be because m is a prime number.

Section 2.12.

Page 57.

> **Problems**
>
> Using the above method, find the GCD in the following cases. Indicate which numbers are relative primes, if any. (For Problem 7, wait and read the next frame before tackling it.)
>
>   1. 14 and 33. 2. 35 and 45. 3. 16 and 35. 4. 48 and 84.
>   5. 372, 270, and 186. 6. 128, 129, and 130. 7. 536 and 642.
>   8. Mom and Son are making party favor bags. They have 18 toy cars, 30 assorted pencils, and 42 chocolates. If they want to give the same party favors to all, what is the biggest number of guests they can accommodate, with no party favor remaining?

SOLUTIONS

In each case, we use the prime factorization as an argument.

1. 14 and 33 are relative primes: $14 = 2 \times 7$ and $33 = 3 \times 11$.

2. $GCD(35, 45) = 5$. We have $35 = 5 \times 7$ and $45 = 3 \times 3 \times 5$.

3. 16 and 35 are relative primes because $16 = 2^4$ and 35 is an odd number.

4. $GCD(48, 84) = 12$, since $48 = 2 \times 2 \times 2 \times 2 \times 3$ and $84 = 2 \times 2 \times 3 \times 7$.

5. $GCD(372, 270, 186) = 6$.
$372 = 2 \times 2 \times 3 \times 31$,
$270 = 2 \times 3 \times 3 \times 3 \times 5$,
$186 = 2 \times 3 \times 31$.

6. 128, 129, and 130 are relative primes because $128 = 2^7$ and 129 is an odd number.

7. $642 - 536 = 106 = 2 \times 53$. We have $GCD(642, 536) = 2$ because neither 642 nor 536 has 53 as a prime factor.

8. The greatest number of guests is the GCD of 18, 30 and 42. Since we have $18 = 2 \times 3 \times 3$, $30 = 2 \times 3 \times 5$, and $42 = 2 \times 3 \times 7$. The greatest number of guests is $2 \times 3 = 6$. Each guest will receive 3 pencils, 5 toy cars, and 7 chocolates.

Page 58.

Problems

Use Rule [GCDif] to compute the GCD of the three pairs of numbers:

1. 743 and 503.   2. 421 and 223.   3. 1523 and 1447.

SOLUTIONS

In each of the three problems, we have two relative primes.

1. We have $743 - 503 = 240 = 2^4 \times 3 \times 5$, and neither 743 nor 503 is divisible by 2 or 3, or 5.

2. $421 - 223 = 198 = 2 \times 3 \times 3 \times 11$, and neither 421 nor 223 is divisible by 2 or 3, or 11.

3. $1523 - 1447 = 76 = 2 \times 2 \times 19$, and neither 1523 nor 1447 is divisible by 2 or 19.

Section 2.13.

Page 60.

Problems

Using the [LCMratio] method, compute the LCM in the three cases:

  1. $5,971$ and $6,377$.   2. $4,569$ and $4,803$.   3. 985 and $1,115$.

Solve the word problems.

  4. Sam must choose a number between 55 and 101 that is a multiple of 3, 5, and 10. Find out all the numbers that he can choose.

  5. Lily must choose a number between 67 and 113 that is a multiple of 3, 5, and 9. Find out all the numbers that she can choose.

  6. Frank must choose a number between 49 and 95 that is a multiple of 3, 4, and 9. Find out all the numbers that he can choose.

  7. Hamburger patties come in packages of 6, and hamburger buns in bags of 8. A party hostess wants to avoid any left over buns or patties. At least how many burgers will she have to make to ensure that there are no left over buns without patties and vice versa?

SOLUTIONS

1. Using the [LCMratio] method, we obtain, with 911 and 853 being prime numbers:
$$\underbrace{6377}_{911\times 7} - \underbrace{5971}_{853\times 7} = 406 = 2\times 7\times 29,$$

   So 7 is the GCD of 6,377 and 5,971. This implies
$$LCM(5971, 6377) = \tfrac{5971\times 6377}{7} = 5,439,581.$$

   (By the way, we sometimes omit commas in writing numbers like 5971 or 6377. Writing $LCM(5,971,6,377)$ would be ambiguous.)

2. $\underbrace{4803}_{1523\times 3} - \underbrace{4569}_{1601\times 3} = 234 = 13\times 3\times 3\times 2$, with 1523 and 1601 being prime numbers. So, 3 is the GCD of 4803 and 4569. This implies
$$LCM(4803, 4569) = \frac{4803\times 4569}{3} = 7,314,969.$$

3. $\underbrace{1115}_{223\times 5} - \underbrace{985}_{197\times 5} = 130 = 13\times 5\times 2$, with 223 and 197 being prime numbers.

   So, 5 is the GCD of 1115 and 985. This implies
$$LCM(1115, 985) = \frac{1115\times 985}{5} = 219,655.$$

4. To solve Sam's problem, we could built the table of the multiples of 3, 5, and 10, and stop when we find the common multiples of 3, 5 and 10 that are between 55 and 101. But there is a quicker method. Since 10 is a multiple of 5, we only have to consider 3 and 10. Their product 30 is the smallest common multiple of 3, 5, and 10, but is not between 55 and 101. The next common multiple is 60, and the next one is 90. Both are between 55 and 101. Any other common multiple is above 101. So, the answer to Sam's problem is 60 and 90.

5. The only number is 90. We have $67 < 90 < 113$.

6. Here also, there is only one number: 72. We have $49 < 72 < 95$.

7. $24 = 4 \times 6 = 3 \times 8$: four packages of hamburgers, and three bags of buns.

Page 61.

> **A problem for the very curious students.** Prove that the Rule for Computing the Least Common Multiple (LCM) of two natural numbers m and m is correct. The LCM Rule is
>
> [LCMratio] $\qquad LCM(n,m) = \dfrac{n \times m}{GCD(n,m)}. \qquad (n \neq m).$

SOLUTION.

Suppose that two natural numbers m and n have the prime factorization:

$$m = \underbrace{m_1 \times \ldots \times m_i}_{m_*} \times \underbrace{m_{i+1} \times \ldots \times m_j}_{GCD(m,n)}, \quad n = \underbrace{n_1 \times \ldots \times n_k}_{n_*} \times \underbrace{n_{k+1} \times \ldots \times n_\ell}_{GCD(m,n)}..$$

Then, successively:

$$\frac{m \times n}{GCD(m,n)} = \frac{m_* \times \cancel{GCD(m,n)} \times GCD(m,n) \times n_*}{\cancel{GCD(m,n)}} = LCM(m,n).$$

Section 2.14
Pages 63-64

> **Problems**
>
> 1. Without a calculator, compute the sum $1 + 2 + \ldots + 65$.
>
> 2. The formula $2 + 4 + 6 + \ldots + (2n-4) + (2n-2) + 2n = n(n+1)$ gives the sum of the first n even counting numbers. Apply that formula to compute the sum $2 + 4 + 6 + \ldots + 40 + 42 + 44$.

3. Use the figure on the right to explain the formula

$$1 + 3 + \ldots + 21$$
$$= 1 + 3 + \ldots + 2 \times 11 - 1 = 11^2$$

for computing the sum of the first eleven odd numbers.

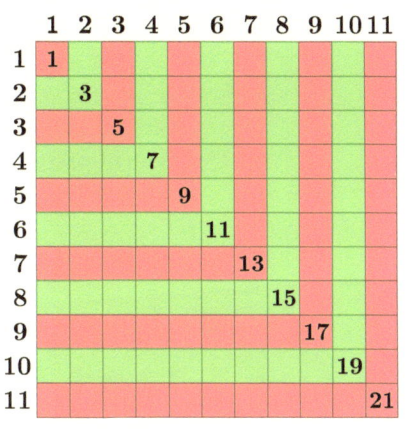

Problems continuation for the very curious students.

4. Prove that the formula

$$2 + 4 + 6 + \ldots + (2n-4) + (2n-2) + 2n = n(n+1) \quad (2.1)$$

is the correct one for computing the sum of the first n even numbers. You can do that directly from the formula itself, or by using the 'picturing the square' method above.

5. On the basis of this case, find the formula for computing the sum of the first n odd numbers, that is, the sum

$$1 + 3 + \ldots + (2n-1).$$

Compute the sum: $1 + 3 + 5 + \ldots + 71 + 73 + 75.$

SOLUTIONS

1. 2145.   2. 506.

3. Looking at the figure makes it cleat that the sum $1 + 3 + 5 + \ldots + 21$ is counting the total number of little squares in the big square, that is $11^2 = 121$:

   1 is counting one red box;   3 is counting three $(1+1+1)$ green boxes;
   5 is counting five $(2+1+2)$ red boxes;… etc.
   21 is counting twenty one $(10+1+10)$ red boxes.

4. We give two proofs of the formula (2.1). One is by induction. If you have to prove that for all natural numbers n, we have

$$F(1) + F(2) + \ldots + F(n) = K(n),$$

you can do this by first showing that $F(1) = K(1)$, and then proving that, for any natural number n, you have $F(n+1) = K(n) + F(n)$. Applying this type of proof to Formula (2.1), we get the following. We have $2 = 1 \times 2 = 1 \times (1+1) = 2$. Trying also the next one for good measure: $2 + 4 = 1 \times 2 + 2 \times 2 = 2 \times (2+1) = 6$. Suppose that $2 + 4 + \ldots + 2(n-1) = (n-1) \times n$. Then $(n-1) \times n + 2n = n(n+1)$.

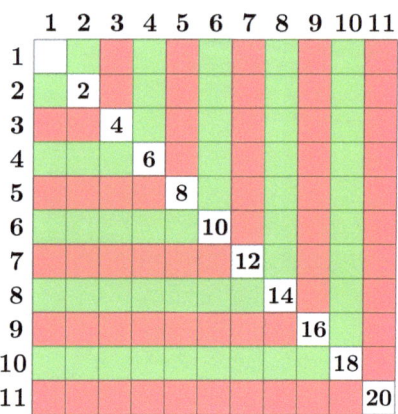

For the second proof, look at the picture on the right. It is clear that $2 + 4 + \ldots + 20$ is equal to the total number 11^2 of squares, less the number of squares in the diagonal, that is 11. So, we get:
$11^2 - 11 = 110 = 10 \times (10+1)$.
In general, with a big square of size n^2 we would get $n^2 - n = (n-1)n$ for the number of little squares outside the diagonal. This argument would work no matter how large that table is.

5. Note that, for any natural number n, we have

$$1 + 3 + 5 + \ldots + (2n-1) = 2 + 4 + 6 + \ldots + 2n - n.$$

For example: $1 + 3 + 5 + 7 = 2 + 4 + 6 + 8 - 4$.

So, we get the formula:

$$1 + 3 + \ldots + (2n-1) = n(n+1) - n = n^2, \quad \text{which gives in particular:}$$

$$1 + 3 + \ldots + 75 = 1 + 3 + \ldots + (2 \times 38 - 1) = 38^2 = 1444.$$

Page 65.

Problems

Suppose that Emile has joined our four athletes. So, we now have five athletes to rank.

1. How many possible rankings of the five athletes are there in which Emile occupies the second rank?
2. In which Emile is in the second rank and Augustus in the fourth?
3. In which Emile occupies an even rank (that is, rank 2 or rank 4)?
4. In which Emile is ranked before Augustus?

SOLUTIONS

1. 24.   2. 6.   3. 48.   4. 60.

Page 67.

> **For the very curious students.** For his war against the Saxons, King Arthur has chosen the following five knights of the Round Table as leaders of his army:
>
> Anselm, Bors, Cador, Dagonel and Erec (for short $a, b, c, d,$ and e).
>
> Arthur must rank these knights according to their valor, and then select the supreme commander. It was resolved that, before their departure, each of the knights must marry one of the following five maids of Queen Guinevere:
>
> Alice, Beatrice, Cecily, Diana, and Emma (for short $a', b', c', d',$ and e').
>
> Queen Guinevere also ranked the maids, according to their beauty. For some reason, King Arthur and Queen Guinevere decided that each knight must marry a maid of equal ranking. However, they did not know that Anselm and Alice were secretly in love, and so were Bors and Beatrice, Cador and Cecily, Dagonel and Diana, and Erec and Emma.
>
> **Problems**
>
> 1. How many ordered pairs of rankings are there? An example of an ordered pair of rankings is $(bcdae, a'b'e'c'd')$.
>
> 2. In how many ordered pairs of rankings do a and a' have the same rank? One example is $(caedb, d'a'b'e'c')$.
>
> 3. In how many pairs of rankings is the rank of a above that of a'?
>
> 4. In how many pairs of rankings will each of the knights be able to marry his secret love?

SOLUTIONS

1. 14,400.   2. 2,880.   3. 5,760.   4. 120.

Pages 68-69.

> **Problems**
>
> 1. How many possible basketball teams can Coach Gustav make with 7 players?

2. An earthquake took place in a remote island in the Pacific. An international health organization decides to send teams of nurses and doctors to provide medical help. Fifteen doctors and eight nurses are available. A typical team is made of 3 doctors and 5 nurses. The organization wants to send the best team right away. This best team is one among how many possible teams? (The teams can overlap.)

3. The Smiths are selecting a furniture set. A furniture set has a bed, a desk, and a dresser. There are 3 beds, 3 desks, and 1 dresser to select from. How many different furniture sets could they select?

4. The lunch special at Mai's Restaurant is a sandwich, a drink, and a dessert. There are 2 sandwiches, 4 drinks, and 6 desserts to select from. How many lunch specials are possible?

5. Lewis High School is going to select a committee. The committee will have a faculty member, a male student, a female student, and a parent. Here are the positions and the people interested in each:

    (a) *Faculty Member*: Dr. Carter, Mr. Johnson, Mrs. Bell
    (b) *Male Student*: John, Eric, Jose, Deon, Aldo
    (c) *Female Student*: Ann, Latoya, Sue, Rita, Mary
    (d) *Parent*: Mrs. Aoki, Ms. Martinez, Dr. Hernandez, Mr. Pham, Ms. Chang

    Based on this list, how many ways are there to fill the four committee positions?

6. Felipe has been given a list of 5 bands and asked to cast a vote. His vote must have the names of his favorite and second favorite bands from the list. How many different votes are possible?

7. In a dog show, there will be a gold, a silver, and a bronze medal awarded. There are 4 finalists. In how many ways can the 3 medals be awarded?

8. There will be 4 floats in a parade. The parade organizers are trying to determine the order in which the floats should appear. How many different orders are possible?

9. **For the very curious students.** Coach Gustav has moved to another college and has now 39 players to choose from. How many different basketball matches can he organize? In other words, how many different pairs of combinations of five players, with no common players, can be set up? (To do this problem, you don't need to have a calculator allowing you to compute $x!$ for a natural number x.)

SOLUTIONS

1. 21.   2. 25,480.   3. 9.   4. 48.   5. 375.
6. 10.   7. 24.   8. 24.
9. Simplify and then compute: $\dfrac{39!}{29!\,10!} \times \dfrac{10!}{5!\,5!} = \dfrac{39!}{5!\,5!\,29!}$.

Page 70

We recall that $_nC_k$ is the number of combinations of size k in a collection of size n, with $k \leq n$. The formula for computing $_nC_k$ is

$$_nC_k = \frac{n!}{k!(n-k)!}. \tag{2.2}$$

We have thus

$$_5C_2 = \frac{5!}{2!\,3!} = \frac{5\times 4\times 3\times 2\times 1}{(3\times 2\times 1)\times(2\times 1)} = \frac{5\times 4\times \cancel{3}\times \cancel{2}}{\cancel{3}\times \cancel{2}\times 2} = 10.$$

10. **A problem for the very curious students.** In a school district, thirty-five students have been nominated for a prize. The superintendent must select three of them to receive the prize. She must also rank these three students because the prizes are different: there is a gold medal, a silver medal and a bronze medal. How many possibilities are there? That is, how many ranked combinations of size 3 exists in a collection of size 35?

Solution. Plus a proof that Formula (2.2) is correct.

We know that the number of combinations is $_{35}C_3$. Each of these combination may be ordered in $3!$ ways. So, the total number of possibilities for the superintendent is equal to
$$_{35}C_3 \times 3! \tag{2.3}$$

But there is another way of computing the number of possibilities, which is based on building a tree diagram of the kind displayed on page 66 of Volume 1. (See the next page.)

There are 35 possibilities for the choice of the gold medal student. Having chosen that student, 34 possibilities remain for the choice of the silver medal student, and then 33 for the choice the bronze medal student. So the total number of possibilities is
$$35 \times 34 \times 33$$
which can be rewritten as
$$\frac{35!}{(35-3)!}. \tag{2.4}$$

As the two numbers in (2.3) and (2.4) must be the same, we get
$$_{35}C_3 \times 3! = \frac{35!}{(35-3)!}.$$

Dividing by 3! on both sides gives
$$_{35}C_3 = \frac{35!}{(35-3)!\,3!}.$$

With $35 = n$ and $3 = k$, this gives
$$_nC_k = \frac{n!}{(n-k)!\,k!},$$

which is Formula (2.2).

Chapter 3

The Integers

Section 3.2.

Page 74.

> **Problems**
>
> Place the following points on the grid of integers:
>
> (i) $(6,1)$;   (ii) $(6,6)$;   (iii) $(6,-6)$;   (iv) $(6,-1)$;
>
> (v) $(1,6)$;   (vi) $(1,1)$;   (vii) $(-6,6)$;   (viii) $(-1,6)$.

SOLUTIONS

Each of the eight points below is pictured by red dots on the grid below

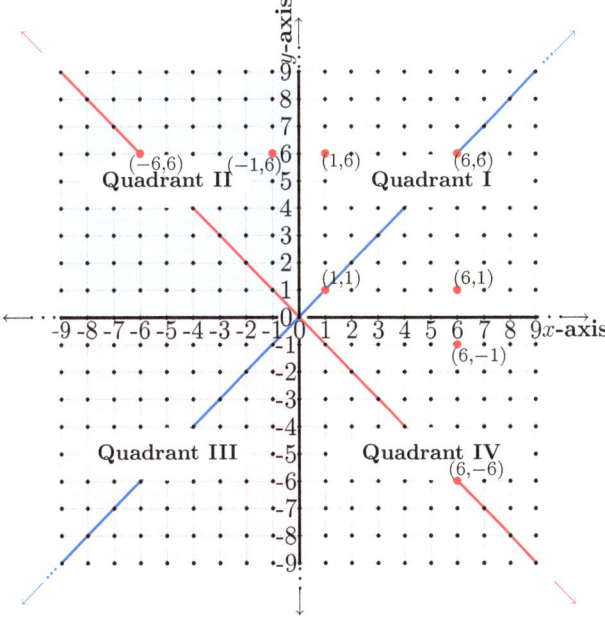

Section 3.3.

Page 77.

> **Problems.** Compute all the expressions below.
> 1. $6 - (-3 - 2)$.
> 2. $-6 - (9 + 6 - 7)$.
> 3. $4 + (-5 - 6) - (-2 - 6)$.
> 4. $-(9 - 5) - (2 - 6)$.
> 5. $13 + (-12 - 5) - (-6 - 8)$.
> 6. $-3 - (13 + 3 - 7) - 4$.
> 7. $-6 + 6$.
> 8. $-5 + (-4)$.
> 9. $-3 + (-5)$.
> 10. $3 + (-5)$.
> 11. $-46 + (-32)$.
> 12. $-25 - 36$.

SOLUTIONS

1. $6 - (-3 - 2) = 11$.
2. $-6 - (9 + 6 - 7) = -14$.
3. $4 + (-5 - 6) - (-2 - 6) = 1$.
4. $-(9 - 5) - (2 - 6) = 0$.
5. $13 + (-12 - 5) - (-6 - 8) = 10$.
6. $-3 - (13 + 3 - 7) - 4 = -16$.
7. $-6 + 6 = 0$.
8. $-5 + (-4) = -9$.
9. $-3 + (-5) = -8$.
10. $3 + (-5) = -2$.
11. $-46 + (-32) = -78$.
12. $-25 - 36 = -61$.

Page 78.

> **Problems**
>
> 1. Eliminate all the parentheses in the order 3, 2, 1 in the example
> $$-6 - \overset{3}{(}\,3 - \overset{2}{(}\,1 - \overset{1}{(}\,2 - 5\,\overset{1}{)}\,\overset{2}{)}\,\overset{3}{)}.$$
> Check that the final result is the same as above.
>
> 2. In the following examples, eliminate the parentheses in two different ways and check that the results are the same.
>   (i) $9 - (2 - (6 + 4) - (5 - 2))$;
>   (ii) $7 - ((-3 + 1) - (2 - 6))$;
>   (iii) $-((3 - 4) + 5 - (6 - 1))$;
>   (iv) $1 - (1 - (1 - (1 - (1 - 2))))$.

SOLUTIONS

1. $-6 - (3 - (1 - (2 - 5))) = -6 - 3 + 1 - 2 + 5 = -5$

2. (i) $9 - (2 - (6 + 4) - (5 - 2)) = 9 - 2 + 6 + 4 + 5 - 2 = 20$;
   (ii) $7 - ((-3 + 1) - (2 - 6)) = 7 + 3 - 1 + 2 - 6 = 5$;
   (iii) $-((3 - 4) + 5 - (6 - 1)) = -3 + 4 - 5 + 6 - 1 = 1$;
   (iv) $1 - (1 - (1 - (1 - (1 - 2)))) = 1 - 1 + 1 - 1 + 1 - 2 = -1$.

Section 3.4.

Page 80.

> **Problems**
>
> 1. Evaluate the following expressions involving absolute value:
>   - (a) $|6|$;
>   - (b) $|-8|$;
>   - (c) $||-10||$;
>   - (d) $|21 - 35|$;
>   - (e) $|-13| - |10 - 14|$;
>   - (f) $||6 - 9| - |-2||$;
>   - (g) $||13| - |8 - 5||$;
>   - (h) $|11 - 13| - |-4|$;
>   - (i) $||16 - 21| - |-7||$.
>
> 2. What is the distance beween 15 and -2?
>
> 3. Show that the following formulas are true:
>   - (a) $|2 - 5| = |5 - 2|$;
>   - (b) $|11 - 6| = |6 - 11|$;
>   - (c) $|5 - 9| \leq |5 - 7| + |7 - 9|$;
>   - (d) $|13 - 10| \leq |13 - 1| + |1 - 10|$.
>
> Problem 3(c) and (d) are two cases of Property [D3], the so-called *triangle inequality* stated below:
> $$|m - n| \leq |m - k| + |k - n|.$$

SOLUTIONS

1. (a) $|6| = 6$;
   (b) $|-8| = 8$;
   (c) $||-10|| = 10$;
   (d) $|21 - 35| = 14$;
   (e) $|-13| - |10 - 14| = 9$;
   (f) $||6 - 9| - |-2|| = 1$;
   (g) $||13| - |8 - 5|| = 10$;
   (h) $|11 - 13| - |-4| = -2$;
   (i) $||16 - 21| - |-7|| = 2$.

2. The distance beween 15 and -2 is 17.

3. (a) $|2 - 5| = |-3| = 3 = |5 - 2|$;
   (b) $|11 - 6| = |5| = |-5| = |6 - 11|$;
   (c) $|5 - 9| = 4 \leq |5 - 7| + |7 - 9| = 2 + 2 = 4$;
   (d) $|13 - 10| = 3 \leq |13 - 1| + |1 - 10| = 12 + 9 = 21$.

Page 83 for the curious student.

The problems refer to a city-block distance map reproduced on the next page.

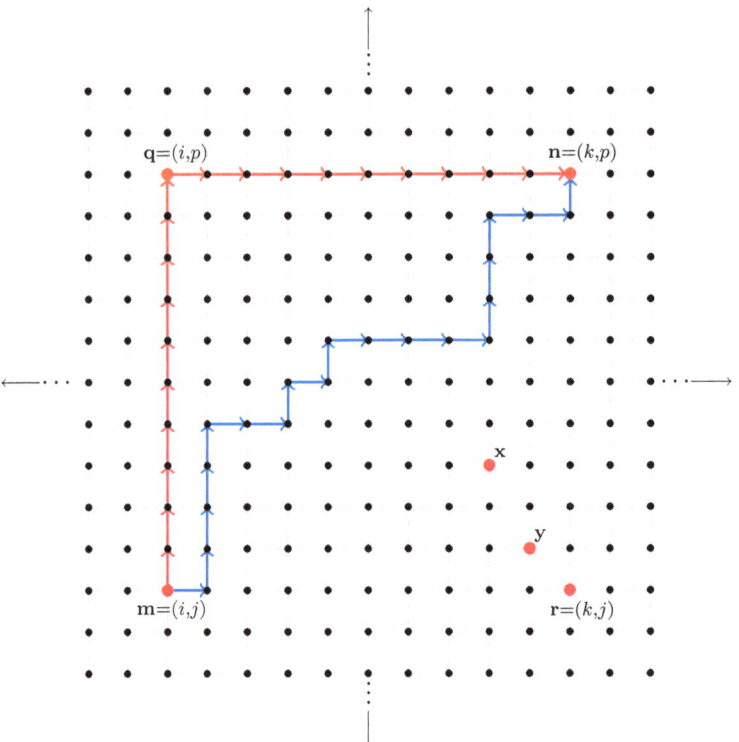

Problems

1. Compute the distance between r and x.

2. Is there a shortest route from m to n passing through x and y? What are conditions for two points ensuring that there is a shortest route from m to n passing through these two points?

3. Is it true that for the point x in the above grid, we have
$$d(m, x) + d(x, n) = d(n, m)?$$

4. Show that the city-block distance satisfies the three properties [D1], [D2] and [D3] listed on page ??. That is, show that, for any points m, n and q on the integer grid, we have
   (a) $d(m, m) = 0$;
   (b) $d(m, n) = d(n, m)$;
   (c) $d(m, n) \leq d(m, q) + d(q, n)$.

Solutions

1. The (city-block) distance between r and x is: $d(r,x) = 5$.

2. There is no shortest route from m to n passing through x and y. For any point p on the grid, there is a shortest route from m to n passing through p if p is one of the points of the 'rectangle' of points of the grid cornered by m and n. Both x and y satisfy that condition. But no shortest route from m to n passing through x (respectively y) also passes through y (respectively x).

   For any two points z and w of the grid, write $R(z,w)$ the rectangle of points of the grid cornered by z and w [1]. (For example, both x and y are in $R(m,n)$.) So, the condition for two points p and q of the grid to be both on a shortest route from m to n is as follows.

   (a) Either p is in $R(m,n)$ and q is in $R(p,m)$;

   (b) or q is in $R(m,n)$ and p is in $R(q,m)$.

3. Yes. we have $\underbrace{d(m,n)}_{20} = \underbrace{d(m,x)}_{11} + \underbrace{d(x,n)}_{9}$.

4. (a) $d(m,m) = 0$: the length of the shortest route from m is 0.

   (b) $d(m,n) = d(n,m)$: any shortest route from m to n can be reversed.

   (c) $d(m,n) \leq d(m,q) + d(q,n)$: either q is on a shortest route from m to n and we have $d(m,n) = d(m,q) + d(q,n)$; or q is not on such a shortest route, and we have $d(m,n) < d(m,q) + d(q,n)$.

Section 3.6.

Page 86. Another problem for the curious students.

We begin by recalling some basic conditions for the addition of integers, which hold for any integers m, n and p.

[C]	Commutativity	$m + n = n + m$
[A]	Associativity	$(m + n) + p = m + (n + p)$
[O]	Opposite	$m + (-m) = 0$
[I]	Identiy	$m + 0 = m$
[S]	Stability	if $m \leq n$ then $m + p \leq n + p$

---

[1] This rectangle may be flat (that is, a line) if z and w are on the same horizontal or vertical route of the grid.

> We show below how the conditions listed on the preceding page can be used to prove that, for any integers m, n and k, we must have
> $$m+n=k \iff n=k-m. \qquad (*)$$
> (Remember: $\Leftrightarrow$ means "if and only if", that is, the left side equality is true whenever the right side equality is true, and the other way around.)
> Here is the proof. With Conditions [S], [C], [O] and [I] as stated earlier:
>
> $$\begin{aligned} m+n=k &\iff m+n+(-m)=k+(-m) &&\text{(by Condition [S])} \\ &\iff m+(-m)+n=k+(-m) &&\text{(by Condition [C])} \\ &\iff 0+n=k+(-m) &&\text{(by Condition [0])} \\ &\iff n=k+(-m) &&\text{(by Condition [I])} \\ &\iff n=k-m &&\text{(Rewriting }+(-m)\text{ as }-m\text{).} \end{aligned}$$
>
> **Problem**
>
> Using the same kind of arguments, prove the following equivalences: for any integers m, n, p and q, we have
> $$m-n=p-q \iff m+q=n+p \iff m-p=n-q.$$
> You only have two equivalences to prove: if you prove
> $$m-n=p-q \Leftrightarrow m+q=n+p \text{ and } m+q=n+p \Leftrightarrow m-p=n-q,$$
> then $\qquad m-n=p-q \iff m-p=n-q$
> follows by the transitivity of the equivalence.

SOLUTION

PROOF THAT: $m-n=p-q \iff m+q=n+p$. Successively,

$$\begin{aligned} m-n=p-q &\iff m-n+q=p-q+q &&\text{(by Condition [S])} \\ &\iff m-n+q=p &&\text{(by Conditions [O] and [I])} \\ &\iff m-n+q+n=p+n &&\text{(by Condition [S])} \\ &\iff m-n+n+q=n+p &&\text{(by Condition [C], applied twice)} \\ &\iff m+q=n+p &&\text{(by Conditions [O] and [I]).} \end{aligned}$$

PROOF THAT: $m+q=n+p \iff m-p=n-q$. Successively,

$$\begin{aligned} m+q=n+p &\iff m+q-q=n+p-q &&\text{(by Condition [S])} \\ &\iff m=n-q+p &&\text{(by Conditions [O] and [C])} \\ &\iff m-p=n-q+p-p &&\text{(by Condition [S])} \\ &\iff m-p=n-q &&\text{(by Condition [O] and [I]).} \end{aligned}$$

The equivalence $m-n=p-q \iff m-p=n-q$ follows by the transitivity of equivalence (c.f. Section 3.5 of Volume I, page 84.).

Section 3.7.

Page 87.

> **Problems**
>
> Compute the following expressions:
>
> 1. $23 - (489 - 980)$.
> 2. $-(35 - 450) - (434 - 567)$.
> 3. $44 - 456 + (85 - 890)$
> 4. $-(487 - 388) - (23 - 678)$
> 5. $-(34 - 67 + 85) - (89 - 99)$
> 6. $-(-44 - (78 - 99))$.

SOLUTIONS

1. 514.   2. 548.   3. -817.
4. 556.   5. -42.   6. 23.

Section 3.8.   **Page 90.**

Rule Symbol	meaning
[M1]	the product of a positive number by a negative number is a negative number
[M2]	the product of a negative number by a negative number is a positive number
[M3]	in an expression without parentheses multiplication precedes addition.

> **Problems**
>
> In each case, identify the properties that you are using to get the result.
>
> 1. Compute the following expressions:
> (i) $-(-73 \times (-67))$;   (ii) $-((-3 \times 6) \times (-8))$;   (iii) $-6 \times (-(-3 + 8))$.
>
> 2. Multiply the following expressions by -8 and compute the result.
> (i) $-(-3 \times (-7))$;   (ii) $-((-6) \times (-3 \times (-8)))$;   (iii) $-(-6 + 8) \times (-2)$.
>
> 3. Multiply the right hand side of each of the inequalities below by some integer so that the two sides become equal.
> (i) $-24 < -3$;   (ii) $-(-3 \times 12) < 6 \times (-2)$;   (iii) $-40 < 5$.
>
> 4. In each case, complete the equation by distributivity.
> (i) $-3 \times (6 - 8) = -3 \times 6 \ldots$;   (ii) $-7 \times (-8 + 9) = -7 \times (-8) \ldots$

SOLUTIONS

1. (i) $\quad -(-73 \times (-67)) = -(73 \times 67)$ [M2]
   $$= -(4891) = -4891;$$

   (ii) $\quad -((-3 \times 6) \times (-8)) = -(-18 \times (-8))$ [M1]
   $$= -(18 \times 8) = -144 \quad \text{[M2]};$$

   (iii) $\quad -6 \times (-(-3+8)) = 6 \times (-3+8)$ [M2]
   $$= 6 \times 5 = 30.$$

2. (i) $\quad -(-3 \times (-7)) \times (-8) = (-3 \times (-7)) \times 8$ [M2]
   $$= 3 \times 7 \times 8 = 168 \quad \text{[M2]};$$

   (ii) $\quad -((-6) \times (-3 \times (-8)) \times (-8) = (-6) \times (-3 \times (-8)) \times 8$ [M2]
   $$= (-6) \times (3 \times 8 \times 8) \quad \text{[M2]}$$
   $$= -(6 \times 3 \times 8 \times 8) \quad \text{[M1]}$$
   $$= -1152;$$

   (iii) $\quad (-(-6+8) \times (-2)) \times (-8) = ((-6+8) \times 2) \times (-8)$ [M2]
   $$= (2 \times 2) \times (-8) = -32 \quad \text{[M1]}$$

3. (i) $\quad -24 = -3 \times 8$ [M1]

   (ii) $\quad -(-3 \times 12) = -(-36) = 36$ [M1]
   $$= \underbrace{(6 \times -2)}_{-12} \times (-3) = 36 \quad \text{[M2]};$$

   (iii) $\quad -40 = 5 \times (-8)$ [M1].

4. (i) $\quad -3 \times (6-8) = -3 \times 6 - (-3) \times (-8)$
   $$= -18 - 24 = -42 \quad \text{[M2]}$$

   (ii) $\quad -7 \times (-8+9) = -7 \times (-8) + (-7 \times 9)$
   $$= 56 + (-63) \quad \text{[M2], [M1]}$$
   $$= 56 - 63 = -7.$$

Section 3.9.

Page 92.

Problems

Give the result of the following divisions, with the remainder if any.

1. $-372 \div 7$. 2. $-345 \div 15$. 3. $-260 \div -11$. 4. $79 \div -8$.

SOLUTIONS

    1. $-372 = -53 \times 7 + 1$;   2. $-345 = -23 \times 15$;
    3. $-260 = -11 \times 23 - 7$;   4. $79 = 8 \times 9 + 7$.

Section 3.10.

Page 94.

> **Problems**
>
> Solve the following equations for the unknowns x, y, z and w:
>
> 1. $2x + 1 = 11$.   2. $2y = 4y - 8$.   3. $-z \times (-3 + 10) = 14$.
> 4. $\frac{w}{3} = 12$.   5. $7z = 2z + 20$.   6. $(1 - x)(10 + 5) = 30$.
> 7. $\frac{w}{3} = 2w - 5$.   8. $2x = 4(3x + 5)$.   9. $3w + 1 = 6 - 2w$.
>
> Note that, to simplify the writing, we have omitted the $\times$ sign in some multiplications. For example, we have written $(1 - x)(10 + 5)$ instead of $(1 - x) \times (10 + 5)$.
>
> All these problems are such that the value of the unknown is an integer.

SOLUTIONS

    1. $x = 5$.   2. $y = 4$.   3. $z = -2$.
    4. $w = 36$.   5. $z = 4$.   6. $x = -1$.
    7. $w = 3$.   8. $x = -2$.   9. $w = 1$.

Pages 102 and 103.

Review problems for Chapter 3.

> 1. Evaluate the following expressions when $c = 9$, $d = 4$, $y = 4$, and $x = 7$.
>
>   (i) $-c + 9d$ (ii) $7y - c$ (iii) $-3c + x$ (iv) $dy - 2x$ (v) $9y + x$
>
> 2. Evaluate the following expressions. Simplify as much as possible.
>
>   (i) $c^2 - 7c + 5$ when $c = 6$ (ii) $m^2 - 6m - 7$ when $m = 2$
>   (iii) $n^2 - 5n - 5$ when $n = -2$ (iv) $a^2 - 7a - 2$ when $a = -6$
>
> 3. Solve for y.
>
>   (i) $y - 4 = 7$ (ii) $9 = 7 + y$ (iii) $3 + y = 6$
>   (iv) $y + 4 = 9$ (v) $y - 9 = 1$ (vi) $y + 3 = 4$

4. Solve for v. Simplify your answer as much as possible.
   - (i) $69 = \frac{v}{3}$
   - (ii) $84 = \frac{v}{4}$
   - (iii) $58 = \frac{v}{2}$
   - (iv) $51 = \frac{v}{3}$
   - (v) $-4 = -\frac{2}{3}v$
   - (vi) $\frac{5}{6}v = -25$
   - (vii) $-10 = -\frac{5}{4}v$
   - (viii) $\frac{3v}{4} = -21$

5. Solve for t. Simplify your answer as much as possible.
   - (i) $18 = -5t + 2(t-3)$
   - (ii) $27 = 5(t+4) + 2t$
   - (iii) $-30 = 7(t-5) - 2t$
   - (iv) $-2(t+5) = 6t + 22$
   - (v) $6(t-3) = 8t + 2$
   - (vi) $5t + 46 = -9(t-2)$

6. Solve for s.
   - (i) $-5(-6s + 3) - 7s = 5(s + 2) - 7$
   - (ii) $2(s + 7) = -7(2s - 4) + 9s$
   - (iii) $-2(4s - 1) + 6s = -2 - 4(2s - 4)$
   - (iv) $2s + 7 + 3(3s + 1) = -3(s + 6)$
   - (v) $4(s - 4) - 29 = -3(4s + 3) - 2s$
   - (vi) $-4(5s - 18) + 8s = 6(s + 6)$

7. Solve the following equations for m in terms of n. For example, in (i), we have $m = \frac{n}{3}$, which is an integer if n is a multiple of 3.
   - (i) $3m = n$
   - (ii) $m - 15 = n$
   - (iii) $15 = \frac{n}{m}$
   - (iv) $m + 7 = n$

For Problems 8, 9 and 10 below, simplify your answers as much as possible:

8. Find the value of $3y - 10$, given that $2y + 7 = 3$.

9. Find the value of $9v + 1$, given that $-2v - 1 = 5$.

10. Find the value of $5w - 4$ given that $4w - 5 = 7$.

11. In each of the three systems below, replace the constant a by a number such that the system has a solution (x, y) which is representable as a point of the integer grid. In other words, the values of x and y are integers.

    - (i) $2x = 3x + 5y - a$
      $4y = x + 20$;
    - (ii) $x = y$
      $x + a = 2y$;
    - (iii) $8x = y + a$
      $2y + 6 = x$.

12. In Problems 11(i), 11(ii) and 11(iii) on page 102, is there more than one possible value of the constant a that would ensure that the system has an integral solution? If you think so, try to formulate, for each of the three cases, a general rule giving all the possible values of a achieving this goal.

13. Solve the following systems of linear equations. That is, find the values of x and y in Problems (i) to (iv), and those of w and z in Problems (v) and (vi). In each case, represent the solutions by points in the integer grid. (So, in Problems (iii), (v) and (vi), you have to relabel w and z as x and y.)

   (i) $6x = 5x + 5y - 3$
   $y = 3x + 23$;

   (ii) $y = 7x + 6$
   $2x = y + 4$;

   (iii) $4x = 3x + y - 1$
   $y = 5x + 5$;

   (iv) $5y = 5x + 5$
   $3x = y + 1$;

   (v) $w = z + 3$
   $3z = 2w - 1$;

   (vi) $z = w + 1$
   $w = 6z + 14$.

14. Expressed in US dollars, Jules' savings increased by 18 is 65. Use the variable c to represent Jules' savings.

15. 108 is the product of Cornelia's age and 6. Use the variable d to represent Cornelia's age.

16. 80 is $\frac{5}{2}$ of Jamon's savings. Use the variable j to represent Jamon's savings.

17. Alfred's age is three times Elen's age. The sum of their ages is 104. What is Elen's age?

18. Victoria received a $70 gift card for a coffee store. She used it for buying some coffee that cost $8 per pound. After buying the coffee, she had $22 left on her card. How many pounds of coffee did she buy?

19. The container that holds the water for the football team is $\frac{1}{4}$ full. After pouring in 5 gallons of water, it is $\frac{1}{2}$ full. How many gallons can the container hold?

20. Aga's gas tank is $\frac{1}{5}$ full. After she buys 13 gallons of gas, it is $\frac{7}{10}$ full. How many gallons can Aga's tank hold?

21. Dina purchased a prepaid phone card for $30. Long distance calls cost 18 cents a minute using this card. Dina used her card only once to make a long distance call. If the remaining credit on her card is $22.08, how many minutes did her call last?

SOLUTIONS

1. (i) 4. (ii) 19. (iii) −20. (iv) 2. (v) 43.
2. (i) −1. (ii) −15. (iii) 9. (iv) 76.
3. (i) $y = 11$. (ii) $y = 2$. (iii) $y = 3$.
   (iv) $y = 5$. (v) $y = 10$. (vi) $y = 1$.
4. (i) $v = 207$. (ii) $v = 336$. (iii) $v = 118$. (iv) $v = 153$.
   (v) $v = 6$. (iv) $v = -30$. (vii) $v = 8$. (viii) $v = 28$.
5. (i) $t = -4$. (ii) $t = 1$. (iii) $t = 1$.
   (iv) $t = 4$. (v) $t = -10$. (vi) $t = -2$.
6. (i) $s = 1$. (ii) $s = 2$. (iii) $s = 2$.
   (iv) $s = -2$. (v) $s = -2$. (vi) $s = 2$.
7. (i) $m = \frac{n}{3}$. (ii) $m = n + 15$. (iii) $m = \frac{n}{15}$. (iv) $m = n - 7$.

8. $3y - 10 = -16$.

9. $9v + 1 = -26$.

10. $5w - 4 = 11$.

11. (i) Solving the system for y, we get $y = \frac{a+20}{9}$. So y is an integer if $a+20$ is a (positive or negative) multiple of 9. If $a = 7$, then $y = \frac{27}{9} = 3$ and $x = 4y - 20 = -8$.

    (ii) Solving the system for y yields $y = a$. So x and y are both integer if a is any integer. We have then $x = y - a$.

    (iii) Solving the system for y gives $y = \frac{a-48}{15}$. So, $a - 48$ must be a (positive or negative) multiple of 15. If $a = 63$, then $y = 1$ and $x = 8$.

12. A problem for the very curious student. We have seen that, in all three cases, there were multiple possible values for the constant a ensuring that the system had an integral solution. The rules for each system were as follows:

    (i) $a + 20$ is a (positive or negative) multiple of 9.

    (ii) a is any integer.

    (iii) $a - 48$ is a (positive or negative) integer.

    For Problem 12 (i) and (iii), the specification "positive or negative" is critical. In Problem (i) for example, if $a = 0$ leads to $9y = 20$. So, y is not an integer. In Problem (ii) however, setting $a = 0$ gives the solution $x = y = 0$.

13. In each of the six cases, the values of the pairs (x, y) solving the system are represented by points on the grid of the next figure on page 43.

(i) The ordered pair (x,y) solving the system is $(-8,-1)$.

(ii) Combining the two equation, we obtain: $2x = 7x + 6 + 4 = 7x + 10$ yielding $-5x = 10$, and so
$$x = -2 \text{ and } y = -8.$$

The ordered pair solving the system is $(-2, -8)$.

For the remaining four cases, we shorten the derivations and drop the comments.

(iii) $\quad x = 5x + 5 - 1$
$\quad\quad -4x = 4$
$\quad\quad x = -1 \text{ and } y = 0.$
$\quad\quad$ Solution: $(-1, 0)$.

(iv) $\quad 3x = x + 2$
$\quad\quad x = 1 \text{ and } y = 2.$
$\quad\quad$ Solution: $(1, 2)$.

For Problems (v) and (vi), we set $w = x$ and $z = y$.

(v) $\quad 2y = 3(y+3) - 1$
$\quad\quad -y = 8$
$\quad\quad x = -5 \text{ and } y = -8.$
$\quad\quad$ Solution: $(-8, -5)$.

(vi) $\quad x = 6(x+1) + 14$
$\quad\quad -5x = 20$
$\quad\quad x = -4 \text{ and } y = -3.$
$\quad\quad$ Solution: $(-4, -3)$.

Placing on the integer grid the six ordered pairs of Problem 13 (i)-(vi): $(-8, -1)$, $(-2, -8)$, $(-1, 0)$, $(1, 2)$, $(8, 5)$, and $(-4, -3)$.

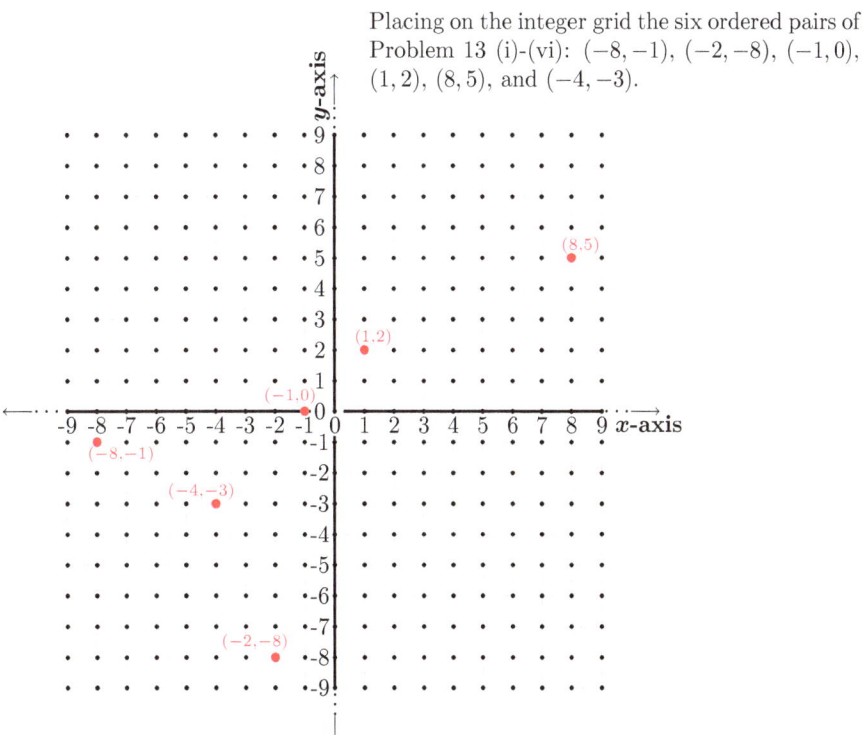

14. Jules' savings are $c = 47$ US dollars: $65 - 18 = 47$.

15. Cornelia is 18 years old. This leads to $108 = 6 \times d$. So, $d = \frac{108}{6}$.

16. We have $80 = j \times \frac{5}{2}$. This implies that $j = \frac{80 \times 2}{5} = 32$.

17. Writing a for Albert's age and e for Elen's age, we have $a + e = 104$ and $3e = a$. This implies $104 = 3e + e = 4e$, yielding $e = \frac{104}{4} = 26$. So, Elen is 26 years old.

18. Writing n for the number of pounds, the problem states that $70 - 22 = n \times 8$. So, we have $n = \frac{48}{8} = 6$. This means that Victoria bought 6 pounds of coffee.

19. We write w for the total number of gallons that the container can hold. The information of the problem can be written in the form: $\frac{1}{4}w + 5 = \frac{1}{2}w$. Multiplying both sides of the equation by 4, we get $4 \times \frac{1}{4}w + 20 = 4 \times \frac{1}{2}w$, that is: $w + 20 = 2w$. Subtracting w on both sides, we get: $w = 20$. So, the container can hold 20 gallons of water.

20. This problem is similar to Problem 19. We use the same kind of reasoning. We write g for the total number of gallons that Aga's gas tank can contain. The information of the problem can be put in the form
$$\frac{1}{5}g + 13 = \frac{7}{10}g.$$
Multiplying by 5 on both sides gives $\quad g + 5 \times 13 = \frac{5 \times 7}{10}g$,
that is $\quad g + 65 = \frac{35}{10}g$.

Multiplying now by 10 on both sides yields $\quad 10g + 650 = 35g$

or $\quad 650 = 25g \quad$ which implies
$$\tfrac{650}{25} = 26 = g.$$
We conclude that Aga's gas tank can contain 26 gallons of gaz. The argument given to solve this problem illustrates a systematic way of presenting a succession of transformations of an equation.

21. In cents, the value of Dina's prepaid phone card is 3000. Let n be the number of minutes of her long distance call. So, the cost of her long distance call is $n \times 18$. Since she has 2208 cents remaining on her card, we have
$$3000 - 2208 = n \times 18.$$
Solving this equation for n gives
$$n = \frac{3000 - 2208}{18} = \frac{792}{18} = 44.$$
So, Dina talked on the phone for 44 minutes.

Chapter 4

The Rational Numbers

Section 4.3.
Page 114.

> **Problems**
>
> We did not yet introduce the order relation $<$ or the addition for the rational numbers. (The order relation is the subject of Section 4.9.) Nevertheless, make good guesses and try to solve the following problems. All of them deal with the 'midpoint' between two rationals.
>
> 1. Show that $\frac{2}{3} < \frac{3}{4}$ and that $\frac{2}{3} < \frac{\frac{2}{3}+\frac{3}{4}}{2} < \frac{3}{4}$.
>
> 2. Show that $-\frac{1}{3} < \frac{2}{5}$ and that $-\frac{1}{3} < \frac{-\frac{1}{3}+\frac{2}{5}}{2} < \frac{2}{5}$.
>
> 3. Show that $-\frac{1}{2} < -\frac{1}{4}$ and that $-\frac{1}{2} < \frac{-\frac{1}{2}+(-\frac{1}{4})}{2} < -\frac{1}{4}$.
>
> 4. The midpoint between 1 and 2 is $\frac{1+2}{2} = 1.5$. The next midpoint, between 1 and 1.5, is 1.25. (You should check it, using the Long Division Algorithm.) Keep computing successively smaller midpoints for a while, as in the graph below. Would you ever reach 1 if you continue indefinitely? This is a problem that puzzled the ancient mathematicians. Its complete solution was worked out around 1820.
>
>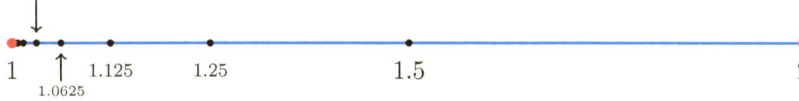
>
> 6. There are 13 books on a shelf. 6 of these books are new. What is the ratio of used books to all books? What is the ratio of used books to new books?
>
> 7. There are 5 boys in a class of 11 students. What is the ratio of girls to all students in the class? What is the ratio of boys to girls?

8. Hans has 7 red marbles and 5 blue marbles. What is the ratio of all marbles to blue marbles? What is the ratio of all marbles to red marbles?

9. Claude has 5 footballs, 10 basketballs, and 10 soccer balls. What is the ratio of soccer balls to all the balls? What is the ratio of soccer balls to balls that are NOT soccer balls?

10. Simplify the fractions:
    (i) $\frac{28}{56}$; (ii) $\frac{24}{6}$; (iii) $\frac{7}{56}$; (iv) $\frac{16}{36}$.

11. For each of the problems below, you have to use $<$, $=$, or $>$ to order the fractions. For example, the correct response to Problem 11(i) is
$$\frac{7}{10} < \frac{8}{10} < \frac{9}{10}.$$

    (i) $\frac{9}{10}, \frac{7}{10}, \frac{8}{10}$.   (ii) $\frac{8}{11}, \frac{3}{11}, \frac{10}{11}$.   (iii) $\frac{6}{7}, \frac{6}{11}, \frac{6}{9}$.
    (iv) $\frac{3}{8}, \frac{3}{11}, \frac{3}{6}$.   (v) $\frac{1}{4}, \frac{2}{7}$.   (vi) $\frac{2}{9}, \frac{1}{5}$.
    (vii) $\frac{2}{9}, \frac{5}{21}$.   (viii) $\frac{9}{20}, \frac{13}{30}$.   (ix) $\frac{6}{7}, \frac{8}{9}$.

SOLUTIONS

1. To show that the inequalities of this problem are true, we make two guesses:

    [G1] If we add the same number on both sides of an inequality, the inequality remains true. We have, for any numbers x, y and z:
    $$x < y \quad \text{if and only if} \quad x + z < y + z.$$
    Note that this statement is an equivalence. Instead of 'if and only if' we could have used '$\iff$', and written
    $$x < y \quad \iff \quad x + z < y + z.$$
    This also applies to our next guess.

    [G2] If we multiply the two sides of an inequality by the same positive number, the inequality remains true. That is, we have for any numbers x, y and $z > 0$:
    $$x < y \quad \iff \quad x \times z < y \times z.$$

    We use these two guesses to solve Problem 1 on page 45, that is, we show that both
    $$\frac{2}{3} < \frac{3}{4} \quad \text{and} \quad \frac{2}{3} < \frac{\frac{2}{3} + \frac{3}{4}}{2} < \frac{3}{4}$$

are true. Starting with $\frac{2}{3} < \frac{3}{4}$, we multiplying both sides by 4. By G2:
$$\frac{2}{3} < \frac{3}{4} \iff \frac{2}{3} \times 4 < \frac{3}{4} \times 4$$
and by simple calculation
$$\iff \frac{2 \times 4}{3} = \frac{8}{3} < \frac{3}{4} \times 4 = 3$$
and summarizing
$$\iff \frac{8}{3} < 3$$
$$\iff \frac{8}{3} \times 3 < 3 \times 3$$
by G2 again, multiplying both sides by 3, and by simplication
$$\iff 8 < 9.$$

So, $\frac{2}{3} < \frac{3}{4}$ is true because it is equivalent to $8 < 9$, which is true.

The argument to prove that $\frac{2}{3} < \frac{\frac{2}{3}+\frac{3}{4}}{2} < \frac{3}{4}$ is similar. We deal with the two inequalities simultaneously. By G2, multiplying the three sides by 2 and rewriting gives,
$$\frac{2}{3} \times 2 = \frac{2}{3} + \frac{2}{3} < \frac{2}{3} + \frac{3}{4} < \frac{3}{4} \times 2 = \frac{3}{4} + \frac{3}{4},$$
and each of the two inequalities simplifies into $\frac{2}{3} < \frac{3}{4}$, which we have just shown to be true.

2.
$$-\frac{1}{3} < \frac{2}{5} \iff -\frac{1}{3} \times 3 < \frac{2}{5} \times 3 \qquad \text{(by G2)}$$
$$\iff -1 < \frac{6}{5} \qquad \text{(simplifying)}$$
$$\iff -1 \times 5 < \frac{6}{5} \times 5 \qquad \text{(by G2)}$$
$$\iff -5 < 6.$$

We use similar steps for the second problem in 2.
$$-\frac{1}{3} < \frac{-\frac{2}{3}+\frac{3}{4}}{2} < \frac{3}{4} \iff -\frac{1}{3} \times 3 < \frac{-\frac{2}{3}+\frac{3}{4}}{2} \times 3 < \frac{3}{4} \times 3 \qquad \text{(by G2)}$$
$$\iff -1 < \frac{\left(-\frac{2}{3}+\frac{3}{4}\right) \times 3}{2} < \frac{9}{4}$$
$$\iff -1 \times 2 < \left(-\frac{2}{3}+\frac{3}{4}\right) \times 3 < \frac{9 \times 2}{4} \qquad \text{(by G2)}$$
$$\iff -2 < .25 < 4.5$$

3. The proofs of the two sets of inequalities use very similar arguments, based on the Rule G2.

4. No. The sequence of midpoints will never reach 1. All the midpoints belong to the half-open interval $]1, \frac{1}{2}[$, a concept that we will study in Chapter 6.

Page 115. Continuation of solutions.

6. Ratio of used books to all books: $\frac{7}{13}$. Ratio of used books to all new books: $\frac{7}{6}$ books.

8. Ratio of all marbles to blue marbles: $\frac{12}{5}$. Ratio of all marbles to red marbles: : $\frac{12}{7}$ books.

10. (i) $\frac{1}{2}$; (ii) 4; (iii) $\frac{1}{8}$; (iv) $\frac{4}{9}$.

11. (i) $\frac{7}{10} < \frac{8}{10} < \frac{9}{10}$. (ii) $\frac{3}{11} < \frac{8}{11} < \frac{10}{11}$. (iii) $\frac{6}{11} < \frac{6}{9} < \frac{6}{7}$.

    (iv) $\frac{3}{11} < \frac{3}{8} < \frac{3}{6}$. (v) $\frac{1}{4} < \frac{2}{7}$. (vi) $\frac{1}{5} < \frac{2}{9}$.

    (vii) $\frac{2}{9} < \frac{5}{21}$. (viii) $\frac{13}{30} < \frac{9}{20}$. (ix) $\frac{6}{7} < \frac{8}{9}$.

Section 4.4.

Page 116.

Problems

In each of the multiplications below, the result should be a repeating or terminating decimal, or an integer.

1. $\frac{47}{-28} \times \frac{-8}{6}$.   2. $\frac{-19}{-8} \times \frac{-3}{6}$.   3. $\frac{-5}{6} \times \frac{7}{-8}$.

4. $\frac{-4}{12} \times \frac{72}{3}$.   5. $\frac{-19}{27} \times \frac{27}{-19}$.

In Problem 5, is there any reason why you should not bother computing anything because the result is immediate?

SOLUTIONS

1. $2.\overline{238095}$.   2. -1.1875.   3. $0.7291\bar{6}$.

4. -8.   5. 1.

In Problem 5, the factors in the numerator and in the denominator are the same.

Page 117.

> **Problems**
>
> Compute the results of the following expressions. In each case, compute the LCM of the denominators, and then rewrite each ratio with the LCM as the common denominator. Note that the numbers in some of the ratios are negative. For example, in Problem 5, we must subtract $\frac{9}{-8}$. This problem should be rewritten as $\frac{20}{-7} - \frac{9}{-8} = -\frac{20}{7} - (-\frac{9}{8}) = -\frac{20}{7} + \frac{9}{8}$.
>
> 1. $\frac{23}{7} - \frac{11}{21}$; 2. $\frac{45}{5} + \frac{9}{7}$; 3. $\frac{789}{18} + \frac{18}{72} - \frac{8}{36}$; 4. $-\frac{31}{10} - \frac{32}{11} - \frac{33}{12}$.
> 5. $\frac{20}{-7} - \frac{9}{-8}$; 6. $\frac{-40}{8} + \frac{8}{5}$; 7. $\frac{78}{3} + \frac{-18}{11}$; 8. $-\frac{3}{-9} - \frac{32}{13} - \frac{30}{11}$.

SOLUTIONS

1. $\frac{23}{7} - \frac{11}{21} = \frac{23 \times 3}{21} - \frac{11}{21} = \frac{69}{21} - \frac{11}{21} = \frac{58}{21}$.

2. $\frac{45}{5} + \frac{9}{7} = \frac{45 \times 7}{35} + \frac{45}{35} = \frac{45 \times 8}{35} = \frac{360}{35} = \frac{72}{7}$.

3. $\frac{789}{18} + \frac{18}{72} - \frac{8}{36} = \frac{789 \times 4}{72} + \frac{18 \times 4}{72} - \frac{8 \times 2}{72} = \frac{3156}{72} + \frac{18}{72} - \frac{16}{72} = \frac{3158}{72} = \frac{1579}{38}$.

4. $-\frac{31}{10} - \frac{32}{11} - \frac{33}{12} = -\frac{31 \times 66}{10 \times 66} - \frac{32 \times 60}{11 \times 60} - \frac{33 \times 55}{12 \times 55} = -\frac{2046}{660} - \frac{1920}{660} - \frac{1815}{660} = -\frac{1927}{220}$.

5. $\frac{20}{-7} - \frac{9}{-8} = \frac{20 \times 8}{-7 \times 8} - \frac{9 \times 7}{-8 \times 7} = \frac{160}{-56} - \frac{63}{-124} \frac{97}{-56} = -\frac{97}{56}$.

6. $\frac{-40}{8} + \frac{8}{5} = \frac{-40 \times 5}{8 \times 5} + \frac{8 \times 8}{5 \times 8} = \frac{-200}{40} + \frac{64}{40} = \frac{-136}{40} = \frac{-34}{10}$.

7. $\frac{78}{3} + \frac{-18}{11} = \frac{78 \times 11}{3 \times 11} + \frac{-18 \times 3}{11 \times 3} = \frac{858}{33} + \frac{-54}{33} = \frac{804}{33} = \frac{268}{11}$.

8. $-\frac{3}{-9} - \frac{32}{13} - \frac{30}{11} = \frac{3 \times 143}{9 \times 143} - \frac{32 \times 99}{13 \times 99} - \frac{30 \times 117}{11 \times 117} = \frac{429}{1287} - \frac{3168}{1287} - \frac{3510}{1287} = -\frac{2083}{429}$.

Page 121.

> **Problems**
>
> Using the above conventions and the rules [**I**], [**C**], [**A**], and [**D**], simplify as much as possible the writing of the formulas 1, 2 and 3 below. You should also eliminate the unnecessary parentheses, if any. There may be more than one way to get a simple result. We begin with an example.

> **Continuation**
>
> By [**C**], [**I**] and the writing convention (which allows us to write mn for $m \times n$), we get
>
> $$-\left(\frac{-2}{4} \times m \times \frac{-4}{6} \times n\right) \times (1 \times k + 7 \times j) = -\left(\frac{-2}{4} \times \frac{-4}{6} mn\right) \times (k + 7j)$$
> $$= -\frac{1}{3} mn(k + 7j)$$
>
> because, by calculation, we have
> $$\frac{-2}{4} \times \frac{-4}{6} = \frac{1}{3}.$$
>
> 1. $\frac{2}{3} \times m \times j \times \frac{3}{4} \times t$.
>
> 2. $\dfrac{4 \times n \times (\frac{2}{3} \times p + 6 \times q)}{2 \times n + 4 \times p \times q}$.
>
> 3. $\left(6 \times \dfrac{7}{8 \times j} \times q\right) \times \left(\dfrac{\frac{j}{6} + \frac{j}{7}}{4 \times p \times q}\right)$.

SOLUTIONS

1. $\frac{1}{2}mjt$. 2. $\frac{\frac{4}{3}np + 12nq}{n + 2pq} = \frac{4np + 36nq}{3n + 6pq}$. 3. $\frac{13}{32p}$.

Section 4.5.

Page 122.

> **Problems**
>
> Carry out the following additions or subtractions:
>
> 1. $3.56 + 56.59 + 19.01$;
> 2. $28.26 + 89.02 + 345.76$;
> 3. $901.09 - 45.87$;
> 4. $0.0124 + 101.124 + 79.01$;
> 5. $103.10 - 99.9$.

SOLUTIONS

1. 79.16 2. 463.04 3. 855.22 4. 180.1464 5. 3.2

Page 123.

> **Problems.** Carry out the following multiplications:
> 1. 7.46×6.59 ; 2. 8.26×981.02 ; 3. 90.11×45.87; 4. 901.09×45.87.

SOLUTIONS

    1. 49.1614   2. 8103.2252   3. 4133.3457   4. 41332.9983

Section 4.6.

Page 126.

> **Problems**
> Carry out the following divisions using the Long Division Algorithm.
> 1. $62.6 \div 1000$.   2. $1.28 \div 100$.   3. $18.2 \div 7$.   4. $39.8 \div 10$.
> 5. $8.34 \div 1000$.   6. $23.4 \div 6$.   7. $2.75 \div 5$.   8. $2.12 \div 4$.
> 9. $\frac{.7345}{6.5}$.   10. $\frac{80.9936}{9.08}$.   11. $\frac{89.9052}{45.87}$.   12. $\frac{707.34}{33.54}$.

SOLUTIONS

In Problem 12, we write $\approx$ to mean '*is approximately equal to.*'

    1. 0.0626   2. 0.0128   3. 2.6   4. 3.98
    5. 0.00834   6. 3.9   7. 0.55   8. 0.53
    9. 0.113   10. 8.92   11. 1.96   12. ≈ 21.0894454

Section 4.7.

Pages 128.

> **Problems**
> In Problems 1 to 6 below, round all the numbers to the fourth decimal digit and compute the resulting expressions. To round 6.591592 to the fourth decimal, replace it by the approximation 6.5916. (Calculator allowed.)
> 1. $8.\overline{18} + 43.8\overline{3}$.   2. $1.017857 + 0.\overline{142857}$.   3. $\pi \times 1.73205$.
> 4. $2.64575 + 3.31662$.   5. $0.\overline{846153} + 1.2\overline{6}$.   6. $2 \times \sqrt{2}$.

Problems

In Problems 7 to 11, find the ceiling and the floor of each the numbers.

7. 8.01.   8. 3.995   9. −17.99.   10. −14.01.   11. −0.78.

12. Estimate 7×444 by computing both 7×400 and 7×500. Are the estimates equally close? If not, which estimate is closer to 7×444?

13. Estimate 8×73 by computing both 8×70 and 8×80. Are the estimates equally close? If not, which estimate is closer to 8×73?

14. Estimate 4×85 by computing both 4×80 and 4×90. Are the estimates equally close? If not, which estimate is closer to 4×85?

15. Estimate 6×536 by computing both 6×500 and 6×600. Are the estimates equally close? If not, which estimate is closer to 6×536?

16. Suppose you choose the division $6300 \div 7$ to estimate $6477 \div 7$. What is your estimate? Is your estimate an underestimate or an overestimate?

17. Suppose you choose the division $720 \div 8$ to estimate $696 \div 8$. What is your estimate? Is your estimate an underestimate or an overestimate?

18. Suppose you want to estimate $6867 \div 9$, and you can choose between the divisions $6300 \div 9$ and $7200 \div 9$. Which division gives a better estimate? What is that estimate? Is it an underestimate or an overestimate?

19. Suppose you want to estimate $2648 \div 5$, and you can choose between the divisions $2500 \div 5$ and $3000 \div 5$. Which division gives a better estimate? What is that estimate? Is it an underestimate or an overestimate?

For Problems 20-25, round each decimal number as indicated.

	Number	Approximation to the nearest		Number	Approximation to the nearest
20.	0.676	hundredth	21.	28.14	tenth
22.	1.31	whole number	23.	25.376	hundredth
24.	0.97	tenth	25.	0.8329	thousandth

SOLUTIONS

1. 52.0151   2. 1.1608   3. 5.4416
4. 5.9624   5. 2.1129   6. 2.8284

12. The best estimate is 7×400. We have $7 \times 400 = 2800$, $7 \times 444 = 3108$
    $7 \times 500 = 3500$, with $|3108 - 2800| = 308$ and $|3500 - 3108| = 392$.

14. The two estimates are equally close: 4×85 is the mean of 4×80 and 4×90. We have $340 = 4 \times 85 = \frac{4 \times 80 + 4 \times 90}{2} = \frac{320 + 360}{2}$.

16. $\frac{6300}{7} = 900$ is an underestimate. We have $\frac{6477}{7} = 925.\overline{285714}$.

18. As $\frac{6867}{9} = 763$, the best estimate is $\frac{7200}{9} = 800$. ($\frac{6300}{9} = 700$.)

	Number	Approximation to the nearest		Number	Approximation to the nearest
20.	0.676	hundreth 0.68	21.	28.14	tenth 28.1
22.	1.31	whole number 1	23.	25.376	hundreth 25.38
24.	0.97	tenth 1	25.	0.8329	thousandth 0.833

Section 4.8.
Page 130.

Problems

Write the improper fractions below in the form of mixed numbers.

1. $\frac{34}{11}$.   2. $\frac{-91}{72}$.   3. $\frac{16}{-9}$.   4. $\frac{11}{9}$.

Compute. Express your answers as mixed numbers in simplest form:

5. $-7\frac{3}{8} + 12\frac{1}{3}$.   6. $-4\frac{1}{8} \times 2\frac{2}{3}$:   7. $-9\frac{3}{8} - 18\frac{2}{3}$.

8. $1\frac{2}{7} + 5\frac{4}{7}$.   9. $8\frac{3}{7} + 3\frac{6}{7}$:   10. $3\frac{3}{5} + 9\frac{4}{5}$.

11. $3\frac{1}{7} - 1\frac{2}{7}$.   12. $7\frac{1}{6} - 1\frac{5}{6}$.   13. $4\frac{1}{8} - 2\frac{1}{3}$.

14. $9\frac{1}{6} - 7\frac{4}{5}$.   15. $1\frac{3}{4} \times \frac{5}{7}$.   16. $3\frac{4}{9} \times 1\frac{1}{2}$.

17. $4\frac{1}{8} \times 2\frac{2}{7}$.   18. $7\frac{1}{5} \times 6\frac{3}{4}$.   19. $2\frac{2}{3} \times 4\frac{2}{5}$.

SOLUTIONS

1. $\frac{34}{11} = 3\frac{1}{11}$.   2. $\frac{-91}{72} = -1\frac{19}{72}$.   3. $\frac{16}{-9} = -1\frac{7}{9}$.   4. $\frac{11}{9} = 1\frac{2}{9}$.

5. $-7\frac{3}{8} + 12\frac{1}{3} = 4\frac{23}{24}$.   6. $-4\frac{1}{8} \times 2\frac{2}{3} = -11$.   7. $-9\frac{3}{8} - 18\frac{2}{3} = -28\frac{1}{24}$.

8. $\frac{2}{7} + 5\frac{4}{7} = 6\frac{6}{7}$.   9. $8\frac{3}{7} + 3\frac{6}{7} = 12\frac{2}{7}$:   10. $3\frac{3}{5} + 9\frac{4}{5} = 13\frac{2}{5}$.

11. $3\frac{1}{7} - 1\frac{2}{7} = 1\frac{6}{7}$.   12. $7\frac{1}{6} - 1\frac{5}{6} = 5\frac{1}{3}$.   13. $4\frac{1}{8} - 2\frac{1}{3} = 1\frac{19}{24}$.

14. $9\frac{1}{6} - 7\frac{4}{5} = 1\frac{11}{30}$.   15. $1\frac{3}{4} \times \frac{5}{7} = 1\frac{1}{4}$.   16. $3\frac{4}{9} \times 1\frac{1}{2} = 5\frac{1}{6}$.

17. $4\frac{1}{8} \times 2\frac{2}{7} = 9\frac{3}{7}$.   18. $7\frac{1}{5} \times 6\frac{3}{4} = 48\frac{3}{5}$.   19. $2\frac{2}{3} \times 4\frac{2}{5} = 11\frac{11}{15}$.

54 CHAPTER 4. THE RATIONAL NUMBERS

Section 4.9.

Page 134.

Problems

In each of the four problems below, find the order of the two rational numbers. Verify the result by multiplying the ratios on each side, using the methods of Case 1 and Case 2 described above. Write each side of the final inequality in decimal form. If needed, use $\approx$ to indicate that the number is an approximation.

1. $\frac{7}{16}$ and $\frac{8}{17}$. 2. $-\frac{15}{6}$ and $\frac{-16}{7}$. 3. $-\frac{5}{3}$ and $\frac{0}{17}$. 4. $\frac{-18}{19}$ and $\frac{-17}{18}$.

Proceed the same way with Problems 5, 6 and 7 below, but then multiply both sides of the inequality that you have found by $-\frac{17}{7}$. Again, write each side of the final inequality in decimal form.

5. $\frac{5}{7}$ and $\frac{9}{17}$. 6. $-\frac{14}{5}$ and $\frac{-19}{7}$. 7. $-\frac{53}{3}$ and $\frac{0}{11}$.

Find the ceilings and the floors of each the following numbers.

8. 8.01. 9. 3.995. 10. -17.99. 11. -14.01. 12. -0.78.

SOLUTIONS

1. $0.4375 = \frac{7}{16} < 0.4706 \approx \frac{8}{17}$. 2. $-\frac{15}{6} = -2.5 < -\frac{16}{7} = -2.\overline{285714}$.

5. $\frac{9}{17} \approx 0.5294 < \frac{5}{7} = 0.\overline{714285}$;

$-\frac{17}{7} \times \frac{5}{7} = -\frac{85}{49} = -1.73469... < -\frac{17}{7} \times \frac{9}{17} = -\frac{9}{7} = -1.\overline{285714}$.

8. $\lfloor 8.01 \rfloor = 8$ 9. $\lfloor 3.995 \rfloor = 3$ 10. $\lfloor -17.99 \rfloor = -18$ 11. $\lfloor -14.01 \rfloor = -15$ 12. $\lfloor -0.78 \rfloor = -1$
$\lceil 8.01 \rceil = 9$ $\lceil 3.995 \rceil = 4$ $\lceil -17.99 \rceil = -17$ $\lceil -14.01 \rceil = -14$ $\lceil -0.78 \rceil = 0$

Page 137.

Problems

In each problem, order the numbers from least to greatest.

1. $7.6, 7.608, 6.4212, 7.68$. 2. $1.826, 2.2072, 2.7, 2.27$.
3. $6.3, 6.1431, 6.104, 6.14$. 4. $7.24, 7.1041, 7.4, 7.208$.
5. $2.604, 2.3041, 2.64, 2.3$. 6. $8.05, 8.1, 8.1022, 8.005$.

SOLUTIONS

1. $6.4212 < 7.6 < 7.608 < 7.68$. 2. $1.826 < 2.2072 < 2.27 < 2.7$.
3. $6.104 < 6.14 < 6.1431 < 6.3$. 4. $7.1041 < 7.208 < 7.24 < 7.4$.
5. $2.3 < 2.3041 < 2.604 < 2.64$. 6. $8.005 < 8.05 < 8.1 < 8.1022$.

Chapter 5

Sets, Relations and Functions

Section 5.1.

Page 144.

> **Problems**
>
> In each of Problem 1, 2, 3, and 4, there are two sets and also some elements. Indicate which of the relations $\in$, $\notin$, and $\subset$ are true for the sets and the elements listed. For example, the correct response to Problem 1 is the list:
>
> $$a \in \{a,d\}, \quad a \in \{a,c,d\}, \quad c \notin \{a,d\}, \quad c \in \{a,c,d\},$$
> $$\text{and} \quad \{a,d\} \subset \{a,c,d\}.$$
>
> 1. a, c, $\{a,d\}$, $\{a,c,d\}$.
>
> 2. x, b, $\{a,b,x,y\}$, $\{y,x\}$.
>
> 3. 5, 7, the set W of all the prime divisors of 63, and the set $\{1,3,5,\ldots\}$ of odd numbers.
>
> 4. a, $\varnothing$, $\{a,b,\varnothing\}$. (This one may be tricky. A set can be an element of another set. Why not?)
>
> 5. Write the set $\{\lfloor 2.56 \rfloor, \lceil 2.21 \rceil\}$ without the floor and ceiling signs.
>
> 6. Find two sets S and T such that
>
> $$\{2,4,6\} \subset S \subset T \subset \{1,2,\ldots,7\}$$
>
>   and moreover, there is a set V such that $S \subset V \subset T$.

SOLUTIONS

2. $x \in \{a,b,x,y\}$, $x \in \{y,x\}$, $b \in \{a,b,x,y\}$, $b \notin \{y,x\}$, $\{y,x\} \subset \{a,b,x,y\}$.

3. $W = \{3,7\}$, $5 \in \{1,3,5,\ldots\}$, $7 \in W \subset \{1,3,5,\ldots\}$, $5 \notin W$.

4. $a \in \{a,b,\varnothing\}$, $a \notin \varnothing$, $\varnothing \in \{a,b,\varnothing\}$, $\varnothing \subset \{a,b,\varnothing\}$. (The argument used on page 142 to prove that $\varnothing \subseteq S$ can be used to prove that $\varnothing \subset \{a,b,\varnothing\}$.)

5. $\{\lfloor 2.56 \rfloor, \lceil 2.21 \rceil\} = \{2,3\}$.

6. $\{2,4,6\} \subset \underbrace{\{1,2,4,6\}}_{S} \subset \underbrace{\{1,2,3,4,6\}}_{V} \subset \underbrace{\{1,2,3,4,5,6\}}_{T} \subset \{1,2,3,4,5,6,7\}$.

Section 5.3.

Page 148.

> **Problems**
>
> 1. In the Venn diagram of Figure 5.1 on page 58, suppose that
>
> $$S = \{a,b,c,e\}, \quad T = \{b,c,d,f\}, \quad \text{and} \quad V = \{c,e,f,g\}. \qquad (5.1)$$
>
>   Give the contents of the sets of the seven regions of the Venn diagram. Compute carefully each of the sets using the definitions of the operations $\cap$, $\cup$ and $\setminus$. For example, we have
>
> $$\begin{aligned} S \setminus (T \cup V) &= S \setminus (\{b,c,d,f\} \cup \{c,e,f,g\}) &&\text{(by the definition of } T \text{ and } V) \\ &= S \setminus \{b,c,d,e,f,g\} &&\text{(by definition of union)} \\ &= \{a,b,c,e\} \setminus \{b,c,d,e,f,g\} &&\text{(by the definition of } S) \\ &= \{a\} &&\text{(by definition of set difference)}. \end{aligned}$$
>
>   Try to imitate this type of argument for the remaining six regions.
>
> 2. Use the Venn diagram of Figure 5.1 to find out the four sets below. Verify your results by replacing the sets S, T and V by their contents given in (5.1).
>
>   (a) $(S \setminus T) \cup V$; show that this set may be different from $S \setminus (T \cup V)$; in particular, when S, T and V have the contents given by (5.1).
>
>   (b) $V \cap (S \cup T)$.
>
>   (c) $(T \cap V) \setminus S$; is this another way of specifying one of the seven regions?
>
>   (d) $((S \cap V) \cup (T \cap V)) \setminus (S \cap T \cap V)$. This turns out to be the union of two regions. Which are these?

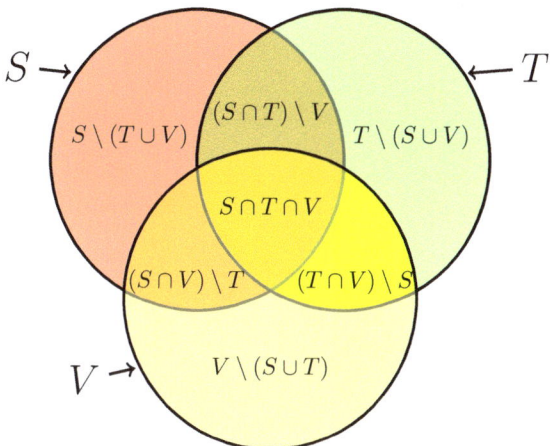

Figure 5.1: The Venn diagram of all the possible regions created by three sets S, T and V. Note the use of parentheses. For example, $S \setminus (T \cup V)$ is not the same set as $(S \setminus T) \cup V$. (See Problem 2 above.)

SOLUTIONS

1.

$$\begin{aligned}
(S \cap T) \setminus V &= (\{a,b,c,e\} \cap \{b,c,d,f\}) \setminus V && \text{(by the definitions of } S \text{ and } T) \\
&= \{b,c\} \setminus V && \text{(by definition of intersection)} \\
&= \{b,c\} \setminus \{c,e,f,g\} && \text{(by the definition of } V) \\
&= \{b\} && \text{(by definition of set difference).}
\end{aligned}$$

$$\begin{aligned}
S \cap T \cap V &= \{a,b,c,e\} \cap \{b,c,d,f\} \cap V && \text{(by the definition of } S \text{ and } T) \\
&= \{b,c\} \cap V && \text{(by definition of intersection)} \\
&= \{b,c\} \cap \{c,e,f,g\} && \text{(by the definition of } V) \\
&= \{c\} && \text{(by definition of intersection).}
\end{aligned}$$

$$T \setminus (S \cup V) = T \setminus (\{a,b,c,e\} \cup \{c,e,f,g\}) \quad \text{(by the definition of } S \text{ and } V\text{)}$$
$$= T \setminus \{a,b,c,e,f,g\} \quad \text{(by definition of union)}$$
$$= \{b,c,d,f\} \setminus \{a,b,c,e,f,g\} \quad \text{(by the definition of } T\text{)}$$
$$= \{d\} \quad \text{(by definition of set difference).}$$

$$(S \cap V) \setminus T = (\{a,b,c,e\} \cap \{c,e,f,g\}) \setminus T \quad \text{(by the definition of } S \text{ and } V\text{)}$$
$$= \{c,e\} \setminus T \quad \text{(by definition of intersection)}$$
$$= \{c,e\} \setminus \{b,c,d,f\} \quad \text{(by the definition of } T\text{)}$$
$$= \{e\} \quad \text{(by definition of set difference).}$$

$$(T \cap V) \setminus S = (\{b,c,d,f\} \cap \{c,e,f,g\}) \setminus S \quad \text{(by the definition of } T \text{ and } V\text{)}$$
$$= \{c,f\} \setminus S \quad \text{(by definition of intersection)}$$
$$= \{c,f\} \setminus \{a,b,c,e\} \quad \text{(by the definition of } S\text{)}$$
$$= \{f\} \quad \text{(by definition of set difference).}$$

$$V \setminus (S \cup T) = V \setminus (\{a,b,c,e\} \cup \{b,c,d,f\}) \quad \text{(by the definition of } S \text{ and } T\text{)}$$
$$= V \setminus \{a,b,c,d,e,f\} \quad \text{(by definition of union)}$$
$$= \{c,e,f,g\} \setminus \{a,b,c,d,e,f\} \quad \text{(by the definition of } V\text{)}$$
$$= \{g\} \quad \text{(by definition of set difference).}$$

2. We omit the justifications of each line.

   (a)
   $$(S \setminus T) \cup V = (\{a,b,c,e\} \setminus \{b,c,d,f\}) \cup V$$
   $$= \{a,e\} \cup V$$
   $$= \{a,e\} \cup \{c,e,f,g\}$$
   $$= \{a,c,e,f,g\}$$
   $$\neq S \setminus (T \cup V) = \{a\}.$$

   (b)
   $$V \cap (S \cup T) = \{c,e,g,f\} \cap (\{a,b,c,e\} \cup \{b,c,d,f\})$$
   $$= \{c,e,g,f\} \cap \{a,b,c,d,e,f\}$$
   $$= \{c,e,f\}.$$

   (c)
   $$(T \cap V) \setminus S = (\{b,c,d,f\} \cap \{c,e,g,f\}) \setminus \{a,b,c,e\}$$
   $$= \{c,f\} \setminus \{a,b,c,e\}$$
   $$= \{f\}.$$

(d) $((S \cap V) \cup (T \cap V)) \setminus (S \cap T \cap V)$
$= ((S \cap V) \cup (T \cap V)) \setminus \{c\}$
$= ((\{a, b, c, e\} \cap \{c, e, f, g\}) \cup (T \cap V)) \setminus \{c\}$
$= (\{c, e\} \cup (T \cap V)) \setminus \{c\}$
$= (\{c, e\} \cup (\{b, c, d, f\} \cap \{c, e, f, g\})) \setminus \{c\}$
$= (\{c, e\} \cup \{c, f\}) \setminus \{c\}$
$= \{e, f\} = ((S \cap V) \setminus T) \cup ((V \cap T) \setminus S).$

Section 5.4.

Page 150.

Problems

In the six problems below, rewrite the set given without using the bar notation. The basic set is $\mathbb{Z} = \{0, 1, -1, 2, -2, \dots\}$, the set of of integers. For example, one way to write the correct response to Problem 1 is

$$\overline{\{1, -1, 2, -2\}} = \{0\} \cup \{3, -3, 4, -4, \dots, n, -n, \dots\}.$$

1. $\overline{\{1, -1, 2, -2\}}$. 2. $\overline{\{-1, -2, \dots, -n, \dots\}}$. 3. $\overline{\mathbb{N}}$.

4. $\overline{\{0, 1, -1\}}$. 5. $\overline{\{2, -2, 4, -4, \dots, 2^n, -2^n, \dots\}}$. 6. $\overline{\mathbb{Z}}$.

In Problem 5, we know what the correct response is: it is the set of all integers which are neither a power of 2, nor the opposite of such a power. But how do write this in mathematics? A special notation to that effect is introduced in Section 5.6 in Volume I. In Problem 6, we ask you what the complement of the basic set is, **with respect to itself**.

SOLUTIONS

The basic set is the set $\mathbb{Z}$ of integers.

2. $\overline{\{-1, -2, \dots, -n, \dots\}} = \{0, 1, 2, \dots\} = \mathbb{N}_0.$

3. $\overline{\mathbb{N}} = \{0, -1, -2, \dots, -n, \dots\}.$

4. $\overline{\{0, 1, -1\}} = \{2, -2, 3, -3, \dots, n, -n, \dots\}.$

6. $\overline{\mathbb{Z}} = \varnothing.$

Section 5.6.

Page 155.

Problems

Recall that we write $\mathbb{Z}$ for the set of integers, $\mathbb{N} = \{1, 2, \ldots\}$ for the set of counting numbers, and $\mathbb{N}_0 = \{0, 1, 2, \ldots\}$ for the set of whole numbers.

In Problems 1 to 4, rewrite the set S by listing its elements between curly brackets. For example, the answer to Problem 1 is: $S = \{-1, 0\}$.

1. $S = \{x \in \mathbb{Z} \mid -2 < x \leq 0\}$.
2. $S = \{x \in \mathbb{Z} \mid x \geq -1\}$.
3. $S = \{\frac{z}{11} \mid z \in \mathbb{Z}, 1 < z \leq 4\}$.
4. $S = \{\frac{x}{7} \mid x \in \mathbb{Z}, 2 < x < 5\}$.

In Problems 5 to 8, we give a set S by listing a few of its elements between curly brackets. We ask that you find a property satisfied by these elements and redefine the set by that property. Because we only give a few of the elements, there may be more than one property linking the elements of a list. It is enough that you find just one such property.

For example, a correct response to Problem 5 is:

$$S = \{x \in \mathbb{N} \mid x = 2^n, \ n \in \mathbb{N}_0\}.$$

5. $S = \{1, 2, 4, 8, 16, \ldots\}$.
6. $S = \{3, 9, 27, 81, 243 \ldots\}$.
7. $S = \{1, 4, 7, 10, 13, \ldots\}$.
8. $S = \{0, .5, 1, 1.5, 2, 2.5, 3, \ldots\}$.

SOLUTIONS

1. $S = \{-1, 0\}$.
2. $S = \{-1, 0, 1, 2, \ldots\}$.
3. $S = \{\frac{2}{11}, \frac{3}{11}, \frac{4}{11}\}$.
4. $S = \{\frac{3}{7}, \frac{4}{7}\}$.
6. $S = \{x \in \mathbb{N} \mid x = 3^n, \ n \in \mathbb{N}_0\}$.
7. $S = \{x \in \mathbb{N} \mid x = 3n + 1, \ n \in \mathbb{N}_0\}$.
8. $S = \{x \in \mathbb{N} \mid x = \frac{n}{2}, \ n \in \mathbb{N}_0\}$.

Section 5.7.

Page 161.

> **Problems**
>
> 1. Write the following relations as sets of ordered pairs. In each case, you must carefully define the relation. The correct answer to Problem (a) is:
>
> $R = \{(m, n) \in \mathbb{N} \times \mathbb{N} \mid m < 10,\ n < 10,\ n = k \times m \text{ for some } k \in \mathbb{N}\}$.
>
>   This is a set-theoretical way of writing: The relation R has domain and range equal to $\mathbb{N}_0$ and is defined by
>
> $mRn \iff m < 10,\ n < 10,\ m \text{ is a divisor of } n$.
>
>   (a) For whole numbers, $m < 10$, $n < 10$, m is a divisor of n.
>
>   (b) For whole numbers, $m < 6$, $n < 6$, the ratio $\frac{m}{n}$ is a whole number.
>
>   (c) For whole numbers, $m < 20$, $n < 20$, $m + 3 < n$.
>
> 2. For each of the following relations, check whether any of the three properties of transitivity, reflexivity, and symmetry is satisfied.
>
>   (a) For whole numbers, $m < 6$, $n < 6$, the ratio $\frac{m}{n}$ is a whole number.
>
>   (b) For whole numbers, $m < 20$, $n < 20$, $m + 3 < n$.
>
>   (c) For whole numbers, m and n, $|m - n| < 3$.

SOLUTIONS

1. (b) $R = \{(m, n) \in \mathbb{N} \times \mathbb{N} \mid m < 6, n < 6, \dfrac{m}{n} \in \mathbb{N}\}$
    $= \{(0, 1), (0, 2), (0, 3), (0, 4), (0, 5), (1, 1), (2, 1), (2, 2), (3, 1), (3, 3),$
    $\qquad\qquad (4, 1), (4, 2), (4, 4), (5, 1), (5, 5)\}$. $(\star)$

   (c) $R = \{(m, n) \in \mathbb{N} \times \mathbb{N} \mid m < 20, n < 20, m + 3 < n\}$
    $= \{(0, 4), (0, 5), \ldots, (0, 19), (1, 5), (1, 6), \ldots, (1, 19),$
    $\qquad\qquad \ldots, (14, 18), (14, 19), (15, 19)\}$. $(\star\star)$

2. (a)   This problem involves the relation R in 1.(b) above. We can use the definition of the second equality $(\star)$. This relation is not transitive, not symmetric, and not reflexive.

not transitive	because	$3R3$, $3R1$, and not $1R3$.
not symmetric	because	$0R1$ and not $1R0$.
not reflexive	because	not $0R0$.

(b) This involves the relation R in 1.(c) above. We use the definition in $(\star\star)$. This relation is transitive, because

$$mRn \text{ and } nRP \Longrightarrow m+3 < n < 20 \text{ and } n+3 < p < 20$$
$$\Longrightarrow m+n+6 < n+p$$
$$\Longrightarrow m+6 < p < 20$$
$$\Longrightarrow m+3 < p < 20$$
$$\Longrightarrow mRp$$

The relation R is:

not symmetric   because $0R4$ and not $4R0$.
not reflexive   because   not $0R0$.

Section 5.8.

Page 167-168.

Problems

Evaluate the given functions at the given values:

1. Let $f(x) = -2x - 1$. Find $f(-3), f(0)$, and $f(5)$.

2. Let $f(x) = 4x - 5$. Find $f(-3), f(0), f(3)$, and $f(4)$.

3. Let $f(x) = 2x^3 + 6$, and let $g(x) = 3x - 1$. Find $f(-2)$ and $g(\frac{-4}{3})$.

4. Let $f(x) = -3x - 3$, and let $g(x) = 4x^2 - x - 3$. Find $f(2)$ and $g(4)$.

5. Let $g(x) = x^2 + 3x$. Find $g(2a)$.

6. Let $f(x) = x - 4$. Find $f(x+4)$.

Given the following relations R, find the domain and range of R. Express your answers using set notation:

7. $R = \{(-2, -2), (0, 7), (-3, 4), (4, -3)\}$

8. $R = \{(5, p), (5, n), (1, m)\}$

9. $R = \{(7, 0), (1, -7), (-4, -7)\}$

10. $R = \{(4, b), (4, d), (0, 8), (8, a)\}$

Continuation.

Determine whether or not the following relations and equations represent functions:

11. $R = \{(-2, -7), (-7, 3), (-7, 0), (-2, 0)\}$

12. $R = \{(3, t), (-2, t), (-2, s), (0, s)\}$

13. $R = \{(-3, -3), (5, 5), (2, 2), (2, 5)\}$

14. $R = \{(2, 1), (-8, 3), (2, 2), (-8, 1)\}$

15. $y = 7x^2 + 3x + 6$

16. $x = \frac{y^2}{5}$

17. $x + y^2 = 2$

18. $2x = y^3$

19. $y = \sqrt{x + 2}$

20. $(y + 4)^2 - 4 = x$

21. Rewrite the following equations as functions. In each case, you should write $f(x)$ instead of y.

    (i) $y + 6x = 3$.   (ii) $2y - 3x = y + 5$.

    (iii) $y + x = 7x - 6$. (iv) $3y + 6x = 3$.

22. Construct the Venn diagram of the relation R in the following cases. Which of these relations is a function? If one is a function, is that function 1-1? In each case, specify the domain and the range of the relation or function.

    (i) aRb, bRc, dRc and dRa.

    (ii) xRx, yRb, zRs, and tRs.

    (iii) xRa, xRs, yRb, yRc, zRd, and eRe.

23. Give an example of a relation which is not a function. That relation must satisfy the following condition: if one reverses the order of the terms in each ordered pair, then the relation becomes a function. So, if the ordered pair (x, y) belongs to the relation, one writes (y, x) instead of (x, y).

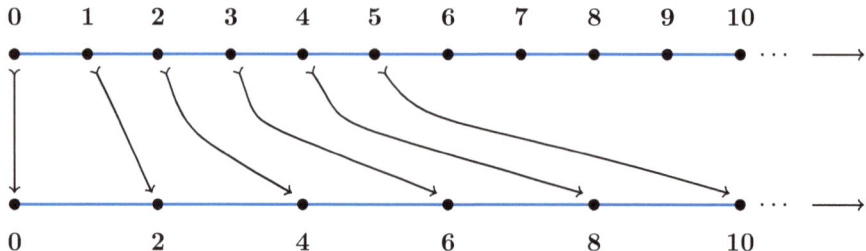

Figure 5.2: The attachment linking whole numbers to even numbers.

Continuation

24. The above figure shows that there is a 1-1 function mapping the set of whole numbers to the set of even whole number. Can you specify what that function is? Show that there is also a 1-1 function from the set of whole numbers to the set of the powers of 2, that is the set $\{1, 2, 4, 8, 16, 32, \ldots\}$. (Remember, we have $2^0 = 1$, $2^1 = 2$, etc.)

---

25. **For the very curious students.**

A 1-1 function also exists, which maps the set of whole numbers to the set of rational numbers. What that says is that there are exactly as many rational numbers as there are whole numbers. This is certainly a more surprising paradox. The proof is not terribly complicate but is still beyond the scope of this book.

SOLUTIONS

1. $f(-3) = 5$, $f(0) = -1$, $f(5) = -11$.
3. $f(-2) = -10$, $g\left(\frac{-3}{4}\right) = -5$.
5. $g(2a) = 4a^2 + 6a$.

For the problems below, we write $\mathcal{D}(R) = \{m \mid mRn \text{ for some n}\}$ for the domain of a relation R, and $\mathcal{R}(R) = \{n \mid mRn \text{ for some m}\}$ for its range.

7. $\mathcal{D}(R) = \{-3, -2, 0, 4\}$, $\mathcal{R}(R) = \{-3, -2, 4, 7\}$
8. $\mathcal{D}(R) = \{1, 5\}$, $\mathcal{R}(R) = \{m, n, p\}$
11.   Not a function: $-7R3$ and $-7R0$.
13.   Not a function: $2R2$ and $2R5$.
15.   A function. This is the equation of a parabola.
17.   A function, for $x \in \,]-\infty, 2]$, with $y = f(x) = \sqrt{2-x}$.
19.   A function, for $x \in [-2, \infty[$, with $y = f(x) = \sqrt{x+2}$.

21. (i) $f(x) = -6x + 3$.   (ii) $f(x) = 3x + 5$.
    (iii) $f(x) = 6x - 6$.   (iv) $f(x) = -2x + 1$.

22. (i) Relation. $\mathcal{D}(R) = \{a, b, d\}$ and $\mathcal{R}(R) = \{b, c, a\}$.
    Not a function: dRc and dRa.

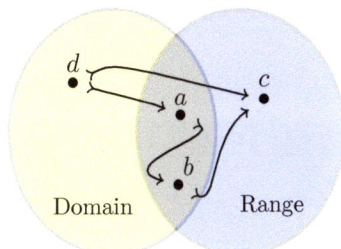

(ii) Function: $\mathcal{D}(R) = \{x, y, z, t\}$ and $\mathcal{R}(D) = \{x, b, s\}$.
Not 1-1: zRs and tRs.

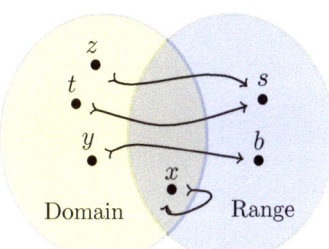

(iii) Relation. $\mathcal{D}(R) = \{x, y, z, e\}$ and $\mathcal{R}(R) = \{a, s, b, c, d, e\}$.

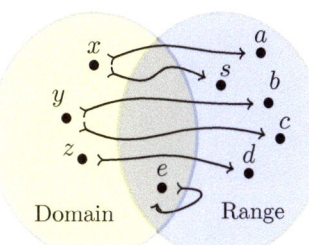

23. $R = \{(a, b), (a, c), (d, e), (d, f)\}$.

24. $f : \mathbb{N}_0 \to \{0, 2, 4, \ldots\} : x \mapsto 2x$; and $g : \mathbb{N} \to \{1, 2, 4, 8, \ldots\} : x \mapsto x^2$; that is,: $f(x) = 2x$ and $g(x) = x^2$.

25. The function k becomes: $k(x) = \begin{cases} 2x + 9 & \text{for } 2 \leq x \leq 6 \\ -x + 27 & \text{for } 6 \leq x \leq 17 \end{cases}$.

Chapter 6

Real Numbers

Section 6.1.

Page 172.

> **Problems**
>
> Use the laws of real numbers to manipulate and simplify the following expressions. You should assume that the letters x, y, u, v and w may represent any real number. No parentheses should remain. For example, the best result for Problem 14 is: $-7u + 4$.
>
> (1) $(y + 5) \times 8$. (2) $2 \times (y - 10)$.
> (3) $(v - 8) \times 5$. (4) $7 \times (7 + v)$.
> (5) $-9 \times (x - 3v - 3)$. (6) $-(5u - w - 1)$.
> (7) $-9 \times (-2x^3 + 4 - 3y^2)$. (8) $(-4 + 2v + 3w) \times (-3)$.
> (9) $-6xy - 2xy$. (10) $-2u + 2u + 4$.
> (11) $4 \times (5y - x) - (6x + 4y)$. (12) $-10u - 2u$.
> (13) $vw - 4v - vw + 3v$. (14) $-2 \times (4u - 2) + u$.
> (15) $-4x^2 + 9x^2$. (16) $-4 \times (4y - u) + 3 \times (6u - 2y)$.
> (17) $(-6w - y) - 4 \times (3y - 4w)$. (18) $x - 6 + x - 6$.
>
> For the six problems below, compute the absolute value of the numbers.
>
> (19) $-\pi$. (20) 3.8.
> (21) -58.1. (22) 23.7.
> (23) 42.42. (24) -5.1601.

SOLUTIONS

1. $8y + 40$
2. $2y - 20$
3. $5v - 40$
4. $49 + 7v$
5. $-9x + 27v + 27$
6. $-5u + w + 1$
7. $18x^3 - 36 + 27y^2$
8. $12 - 6v - 9w$
9. $-8xy$
10. 4
11. $16y - 10x$
12. $-12u$
13. $-v$
14. $-7u + 4$
15. $5x^2$
16. $-22y + 22u$
17. $-13y + 10w$
18. $2x - 12$
19. $|-\pi| = \pi$
20. $|3.8| = 3.8$
21. $|-58.1| = 58.1$
22. $|23.7| = 23.7$
23. $|42.42| = 42.42$
24. $|-5.1601| = 5.1601$

Page 175.

The next 9 problems use the Laws of Exponents. We recall these laws below. The letters x and y represent real numbers, and j, k are any integers.

$$x^1 = x \qquad \text{for any real number } x;$$
$$x^0 = 1 \qquad \text{for any real number } x \neq 0.$$

1. with $x \neq 0$, $\qquad x^j \times x^k = x^{j+k}$;
2. with $x \neq 0$ and $y \neq 0$, $\qquad x^k \times y^k = (xy)^k$.
3. with $x \neq 0$, $\qquad (x^j)^k = x^{j \times k}$.
4. with $x \neq 0$, $\qquad \dfrac{x^j}{x^k} = x^{j-k}$;
5. with $y \neq 0$, $\qquad \dfrac{x^k}{y^k} = \left(\dfrac{x}{y}\right)^k$;
6. with $x \neq 0$, $\qquad x^{-k} = \dfrac{1}{x^k}$.

Problems

In Problems 1 to 8 below, compute the results of each of the expressions, and indicate which of the laws of exponents you have been using. For each expression, you should simplify the formula as much as possible, as we did in the last example. You can then use a calculator for the last computation.

1. $\dfrac{(-6)^{2-5}}{3^{4-6}}$.
2. $\dfrac{(-15.5)^{3-6}}{(3.1)^{4-7}}$.
3. $\dfrac{(-8.2)^{1-4}}{(4.1)^{3-6}}$.
4. $\dfrac{(3^{-2})^6}{3^{-11}}$.
5. $\dfrac{(6.2)^4 \times (6.2)^5}{2^9 \times (3.1)^9}$.
6. $\dfrac{(-6)^6 \times (-3)^6}{2^5 \times 9^5}$.
7. $\dfrac{(7.5)^3}{(4.2)^2} \times \dfrac{(4.2)^3}{(7.5)^2}$.
8. $\dfrac{(-16)^2}{(2^3)^4}$.

9. Try to prove Law 5 for the exponents using the same kind of arguments that we have used to prove Law 4.

SOLUTIONS

1. $\dfrac{(-6)^{2-5}}{3^{4-6}} = \dfrac{(-6)^{-3}}{3^{-2}}$
$= \dfrac{3^2}{(-6)^3}$ (by Exponent Law 6)
$= -\dfrac{1}{24}$ (by computation).

2. $\dfrac{(-15.5)^{3-6}}{(3.1)^{4-7}} = \dfrac{(-15.5)^{-3}}{(3.1)^{-3}}$
$= \left(\dfrac{-15.5}{3.1}\right)^{-3}$ (by Exponent Law 5)
$= (-5)^{-3}$ (by computation)
$= \dfrac{1}{(-5)^3}$ (by Exponent Law 6)
$= -\dfrac{1}{125}$ (by computation).

3. $\dfrac{(-8.2)^{1-4}}{(4.1)^{3-6}} = \dfrac{(-8.2)^{-3}}{(4.1)^{-3}}$
$= \left(\dfrac{-8.2}{4.1}\right)^{-3}$ (by Exponent Law 5)
$= (-2)^{-3}$ (by computation)
$= \dfrac{1}{(-2)^3}$ (by Exponent Law 6)
$= -\dfrac{1}{8}$ (by computation).

4. $\dfrac{\left(3^{-2}\right)^6}{3^{-11}} = \dfrac{3^{-12}}{3^{-11}}$ (by Exponent Law 3)
$= 3^{-1} = \dfrac{1}{3}$ (by Exponent Laws 4 and 6).

9. Proof of Exponent Law 5. Successively, we have, for $x \neq 0$ and $y \neq 0$.

$$\dfrac{x^k}{y^k} = x^k \times \left(\dfrac{1}{y}\right)^k \quad \text{(rewriting)}$$
$$= x^k \times \left(y^{-1}\right)^k \quad \text{(by Law 6)}$$
$$= \left(x \times y^{-1}\right)^k \quad \text{(by Law 2)}$$
$$= \left(x \times \dfrac{1}{y}\right)^k = \left(\dfrac{x}{y}\right)^k \quad \text{(by Law 6 and simplifying).}$$

Page 176.

Problems

For each of the six problems below, you should use the first equation to carry out the substitution in the second equation. Solve then for the variable x using the Substitution Property. More than one substitution may be necessary.

1. $z = \sqrt{9}$
$\dfrac{x}{27} = \dfrac{1}{z}$.

2. $z = y^2 + x$
$y^2 + 3x = z - 6$.

3. $z = (x - y)(x + y)$
$x^2 - y^2 = 2x + z$.

4. $y = 3x^2$
$3(x - x^2 + \dfrac{y}{3}) = 8$.

5. $w = y^2 - 3y$
$(y - 3)y - x = w$.

6. $w = 27$
$(x - 6)(x + 6) = w$.

SOLUTIONS

1. $\begin{cases} z = \sqrt{9} \\ \frac{x}{27} = \frac{1}{z} \end{cases}$
   $\implies \frac{x}{27} = \frac{1}{\sqrt{9}} = \frac{1}{3}$
   $\implies 3x = 27 \implies x = 9$

2. $\begin{cases} z = y^2 + x \\ y^2 + 3x = z - 6 \end{cases}$
   $\implies y^2 + 3x = y^2 + x - 6$
   $\implies 2x = -6 \implies x = -3.$

3. $\begin{cases} z = (x-y)(x+y) \\ x^2 - y^2 = 2x + z \end{cases}$
   $\implies x^2 - y^2 = 2x + (x^2 - y^2)$
   $\implies 0 = 2x$
   $\implies x = 0.$

4. $\begin{cases} y = 3x^2 \\ 3(x - x^2 + \frac{y}{3}) = 8 \end{cases}$
   $\implies 3(x - x^2 + \frac{3x^2}{3}) = 8$
   $\implies 3x = 8$
   $\implies x = \frac{8}{3}.$

5. $\begin{cases} w = y^2 - 3y \\ (y-3)y - x = w \end{cases}$
   $\implies (y-3)y - x = y^2 - 3y$
   $\implies y^2 - 3y - x = y^2 - 3y$
   $\implies -x = 0 \implies x = 0$

6. $\begin{cases} w = 27 \\ (x-6)(x+6) = w \end{cases}$
   $\implies (x-6)(x+6) = 27$
   $\implies x^2 - 36 = 27$
   $\implies x^2 = 63 \implies x = \pm\sqrt{63}$

Section 6.2.

Page 179-180.

Problems

In Problems 1 to 12, evaluate or simplify the expression as much as possible. When applicable, express your answer in lowest terms. You do not need a calculator.

1. $\sqrt{\frac{100}{64}}$. 2. $\sqrt{\frac{48}{27}}$. 3. $\sqrt{\frac{25}{49}}$. 4. $\sqrt{\frac{45}{20}}$.
5. $\sqrt{9y^{10}}$. 6. $\sqrt{16z^6}$. 7. $\sqrt{49x^{14}}$. 8. $\sqrt{81z^{12}}$.
9. $\sqrt{27w^{10}}$. 10. $\sqrt{48z^6}$. 11. $\sqrt{24x^9}$. 12. $\sqrt{18x^{11}}$.

In Problems 13 to 18, calculate both sides and verify that Law 1 or 2 for roots, recalled below is satisfied. **Law 1:** $\sqrt[k]{x} \times \sqrt[k]{y} = \sqrt[k]{xy}$. **Law 2:** $\frac{\sqrt[k]{x}}{\sqrt[k]{y}} = \sqrt[k]{\frac{x}{y}}$. Some of the numbers are repeating decimals. So, pay attention to the rounding of your calculator if you use one.

13. $\sqrt{16.\overline{3}} \times \sqrt{12} = \sqrt{16.\overline{3} \times 12}$.
14. $\sqrt[4]{3} \times \sqrt[4]{27} = \sqrt[4]{3 \times 27}$.
15. $\sqrt[3]{-8} \times \sqrt[3]{27} = \sqrt[3]{-8 \times 27}$.
16. $\sqrt{121} \times \sqrt{196} = \sqrt{121 \times 196}$.
17. $\frac{\sqrt{81}}{\sqrt{-121}} = \sqrt{\frac{81}{-121}}$.
18. $\frac{\sqrt[3]{-804.357}}{\sqrt[3]{29.791}} = \sqrt[3]{\frac{-807.357}{29.791}}$.

SOLUTIONS

1. $\frac{5}{4}$ 2. $\frac{4}{3}$ 3. $\frac{5}{7}$ 4. $\frac{3}{2}$
5. $3y^5$ 6. $4z^3$ 7. $7x^7$ 8. $9z^6$
9. $3w^5\sqrt{3}$ 10. $4z^3\sqrt{3}$ 11. $2x^4\sqrt{6x}$ 12. $3x^5\sqrt{2x}$

The solutions of Problems 13-16 result from Law 1 for roots.

The solutions of Problems 17 and 18 result from Law 2 for roots.

13. $\sqrt{16.\overline{3}} \times \sqrt{12} = \sqrt{16.\overline{3} \times 12}$
    $14 = 14$
14. $\sqrt[4]{3} \times \sqrt[4]{27} = \sqrt[4]{3 \times 27}$
    $3 = 3$
15. $\sqrt[3]{-8} \times \sqrt[3]{27} = \sqrt[3]{-8 \times 27}$
    $-6 = -6$
16. $\sqrt{121} \times \sqrt{196} = \sqrt{121 \times 196}$
    $154 = 154$
17. $\frac{\sqrt{81}}{\sqrt{-121}} = \sqrt{\frac{81}{-121}}$
    no solution

Section 6.3.

Page 184.

Problems

Find the value of x in the 18 equations on the next page. In each case, you should rewrite the equation so that both sides have the same unit. Then, drop the unit and solve the equation for x as usual. For example, the correct response to Problem 10 is $x = \frac{1}{9}$ because we can rewrite that equation as 4 in $= x \times 36$ in. Dropping the 'in', we get $4 = x \times 36$, and so $x = \frac{4}{36} = \frac{1}{9}$.

1. 2.5 cm = x mm. 2. 85 dm = x m. 3. 854 cm = x km.
4. 3.5 m = x km. 5. .05 m = x mm. 6. 9,114 mm = x m.
7. $\frac{1}{8}$ m = x cm. 8. 1.2 m = x dm. 9. 3.7 m = x km.
10. 4 in = x yd. 11. 3.6 ft = $3x$ yd. 12. 1320 ft = x mile.
13. 6 in = x cm. 14. 1.5 ft = $8x$ m. 15. 1500 yd = x km.
16. 2.5 mile = $2x$ km. 17. 6 ft = x m. 18. 40 in = x m.

SOLUTIONS
1. $x = 25$. 2. $x = 8.5$. 3. $x = .00854$.
4. $x = .0035$. 5. $x = 50$. 6. $x = 9.114$.
7. $x = 12.5$. 8. $x = 12$. 9. $x = .0037$.
10. $x = \frac{1}{9}$. 11. $x = .4$. 12. $x = .25$.
13. $x = 15.24$. 14. $x = .05715$. 15. $x = 1.3716$.
16. $x = 2.01168$. 17. $x = 1.8288$. 18. $x = 1.016$.

Page 187.

Problems

Solve each problem by the same method that you used for the length problems. That is, rewrite the equation so that the same unit is used on both sides, drop the unit, and solve for x. In the last line, we use $\approx$ to mean 'approximately equal to.' Still, you should treat such formulas as equations. For the meanings of the abbreviations, see the Appendix on pages 143-146 at the end of this volume.

1. 2,033 g = x kg. 2. 85 dg = x g. 3. 2.4 kg = $.6x$ t.
4. 36 hg = $2x$ mg. 5. 2.5 dag = $.62x$ hg. 6. 3,500 dg = $.1x$ t.
7. $\frac{4}{5}$ mg = $.5x$ dag. 8. 2,500 cg = x hg. 9. .5 dag = $2x$ cg.
10. 84 oz = $24x$ lb. 11. 222 lb = $.2x$ ton. 12. 3200 oz = $2x$ ton.
13. 6 oz $\approx x$ dg. 14. 1.5 lb $\approx 8x$ kg. 15. 3.5 ton $\approx x$ t.

SOLUTIONS
1. $x = 2.033$. 2. $x = 8.5$. 3. $x = .004$.
4. $x = 1,800,000$. 5. $x = .40323$. 6. $x = .0035$.
7. $x = .00016$. 8. $x = .25$. 9. $x = 250$.
10. $x = .21875$. 11. $x = .555$. 12. $x = .05$.
13. $x \approx 1701$. 14. $x \approx .085125$. 15. $x \approx 3.178$.

Page 189.

> **Problems**
>
> Solve each problem by finding the value of x. In the two last lines, $\approx$ means 'approximately equal to.' Deal with such formulas as if they were equations.
>
> 1. $350 \text{ mL} = 2.5\,x \text{ dL}$.
> 2. $.45 \text{ L} = 3\,x \text{ daL}$.
> 3. $25 \text{ L} = 2\,x \text{ hL}$.
> 4. $.05 \text{ kL} = 5\,x \text{ L}$.
> 5. $30 \text{ cL} = 45\,x \text{ mL}$.
> 6. $50 \text{ daL} = .5\,x \text{ hL}$.
> 7. $35 \text{ dL} = 12\,x \text{ L}$.
> 8. $x \text{ daL} = .6 \text{ hL}$.
> 9. $25 \text{ cL} = 300\,x \text{ mL}$.
> 10. $3 \text{ pt} = 40\,x \text{ fl oz}$.
> 11. $2 \text{ qt} = 40\,x \text{ c}$.
> 12. $400 \text{ fl oz} = .2\,x \text{ gal}$.
> 13. $6 \text{ fl oz} \approx x \text{ cL}$.
> 14. $3 \text{ daL} \approx 2\,x \text{ gal}$.
> 15. $15 \text{ c} \approx 3\,x \text{ L}$.
> 16. $2 \text{ dL} \approx x \text{ pt}$.
> 17. $3 \text{ hL} \approx 30\,x \text{ qt}$.
> 18. $7 \text{ qt} \approx 7\,x \text{ L}$.

SOLUTIONS

1. $x = 1.4$
2. $x = .015$
3. $x = .125$
4. $x = 10$
5. $x = 6\frac{2}{3}$
6. $x = 10$
7. $x = \frac{7}{24}$
8. $x = 10$
9. $x = \frac{5}{6}$
10. $x \approx 1.1825$
11. $x \approx .2$
12. $x \approx 15.873$
13. $x \approx 18$
14. $x \approx 3.9682$
15. $x \approx 1.185$
16. $x \approx .4228$
17. $x \approx 10.15708$
18. $x \approx .946$

Page 191.

> **Problems**
>
> Solve each problem by finding the value of x in the equations.
>
> 1. $.5 \text{ wk} = 2.5\,x \text{ d}$.
> 2. $350 \text{ ms} = 2\,x \text{ min}$.
> 3. $2{,}500 \text{ s} = 3\,x \text{ d}$.
> 4. $3000 \text{ min} = .5\,x \text{ wk}$.
> 5. $400 \text{ ms} = 2\,x \text{ s}$.
> 6. $50 \text{ d} = 3\,x \text{ wk}$.

SOLUTIONS

1. $x = 1.4$
2. $x = .002916\overline{6}$
3. $x \approx .009645$
4. $x \approx .59524$
5. $x = .2$
6. $x \approx 2.38095$

Word problems for measurements

Solve the word problems. Include correct units in your answers:

1. On a long hike, Keisha drank 6 pints of water. How much is this in quarts?

2. Manuel drank 24 fluid ounces of juice. How much is this in cups?

3. A bottle holds 48 fluid ounces of lemonade. How much is this in pints?

4. Over a two-hour time period, a snail moved 72 inches. How far is this in yards?

5. A rectangular floor is 7 yards long and 4 yards wide. What is the area of the floor in square feet? (Hint: the area of a rectangle is the product of the lengths of its two sides.)

6. A rectangular flower bed is 15 feet longs and 12 feet wide. What is the area of the flower bed in square yards?

7. Mai is on the swim team. Each week she swims a total of 6000 meters. How many kilometers does she swim each week?

8. A scientist is working with 18 meters of gold wire. How long is the wire in millimeters?

9. Jocelyn's racehorse weighs 700 kilograms. How much does it weigh in grams?

10. A cereal bar contains 2000 milligrams of protein. How much protein does it contain in grams?

11. Karen finished a race in 9 minutes. How many seconds is this?

12. Leila has to give medicine to her cat every morning for the next 21 days. How many weeks is this?

SOLUTIONS

1. 3 quarts.
2. 3 cups.
3. 3 pints.
4. 2 yards.
5. 252 square feet.
6. 20 square yards.
7. 6 kilometers.
8. 18,000 millimeters.
9. 700,000 grams.
10. 2 grams.
11. 540 seconds.
12. 3 weeks.

Page 192.

> **Problems**
>
> Copy each of the numbers below and underline the significant digits.
>
> 1. 0.5400   2. 35.0010   3. 002,501.00
> 4. 30.600   5. 00.4001   6. 5.5550

SOLUTIONS

1. 0.<u>5400</u>   2. 35.<u>0010</u>   3. 002,<u>501</u>.00
4. <u>30.600</u>   5. 00.<u>4001</u>   6. <u>5.5550</u>

Section 6.5.

Page 198.

> **Problems**
>
> For each problem, use the formula
>
> $$d(A,B) = \sqrt{(x_2 - x_1)^2 + (y_2 - y_1)^2}$$
>
> to compute the distance between the points A and B. (Hint: in each case, the result is a positive integer.)
>
> 1. $A = (1.5, 0)$, $B = (-3.5, 12)$.   2. $A = (-3, 7)$, $B = (4, -17)$.
> 3. $A = (2, 5)$, $B = (-6, -10)$.   4. $A = (2, 5)$, $B = (13, 65)$.

SOLUTIONS

1. $d(A, B) = 13$.   2. $d(A, B) = 25$.
3. $d(A, B) = 17$.   4. $d(A, B) = 61$.

Chapter 7

Linear Equations and Linear Inequalities

Section 7.2.

Page 207.

> **Problems**
>
> 1. For each equation, find an ordered pair (x, y) that is a solution to that equation:
>
>   (i) $x + 6y = 5$; (ii) $5x - y = 5$; (iii) $x + 4y = 8$; (iv) $3x - y = 2$.
>
> 2. For each of the four equations below, find the x-intercept and y-intercept of the line given by that equation. (You have to rewrite the equation in the form $y = ax + b$.)
>
>   (i) $9x - 7y = -8$; (ii) $4x - 5y = -14$;
>   (iii) $4x + 3y + 5 = 0$; (iv) $6x - 3y = -15$.
>
> 3. In each of the four cases below, find the slope of the line that passes through the two points:
>
>   (i) $(4, -6), (7, -6)$; (ii) $(-2, -3), (-7, 5)$;
>   (iii) $(2, -4), (7, 5)$; (iv) $(-4, -5), (-4, 6)$.
>
> 4. Find the slopes of the following lines:
>
>   (i) $2y - 5x = -1$ (ii) $2y - 4x - 1 = 0$;
>   (iii) $y = 3$ (iv) $4x + 2y = 1$.

Problems

5. Based on the information given below, find equations for the lines described. Write the equations in slope-intercept form.

   (i) The line that passes through $(-4, -7)$ and has a slope of -2;
   (ii) The line that passes through $(-6, 6)$ and has a slope of -3;
   (iii) The line that passes through $(5, -6)$ and has a slope of 2;
   (iv) The line that passes through $(9, 4)$ and has a slope of 0.

6. Based on the information or equations given below, graph the lines described. Label the x-intercepts and y-intercepts, when appropriate.

   (i) The line whose x-intercept is -4 and whose y-intercept is -6;
   (ii) The line whose x-intercept is 9 and whose y-intercept is 4;
   (iii) $y = \frac{1}{3}x + 4$;
   (iv) $2x - 5y + 20 = 0$;
   (v) $y = -1$;
   (vi) $x = 7$.

SOLUTIONS

1. (i) $(-1, 1)$;   (ii) $(1, 0)$;   (iii) $(4, 1)$;   (iv) $(0, -2)$.

2. (i) x-intercept: $-\frac{8}{9}$, y-intercept: $\frac{8}{7}$
   (ii) x-intercept: $-\frac{7}{2}$, y-intercept: $\frac{14}{5}$
   (iii) x-intercept: $-\frac{5}{4}$, y-intercept: $-\frac{5}{3}$
   (iv) x-intercept: $-\frac{5}{2}$, y-intercept: 5

3. (i) slope: 0; (ii) slope: $-\frac{8}{5}$; (iii) slope: $\frac{9}{5}$; (iv) slope: undefined.

4. (i) slope: $\frac{5}{2}$; (ii) slope: 2; (iii) slope: 0; (iv) slope: -2.

5. (i) $y = -2x - 15$;   (ii) $y = -3x - 12$;
   (iii) $y = 2x - 16$;   (iv) $y = 4$.

6.

(i)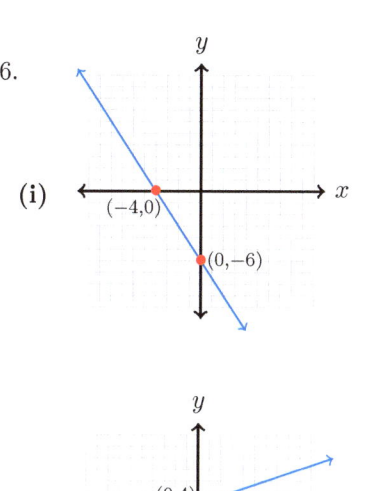

(ii)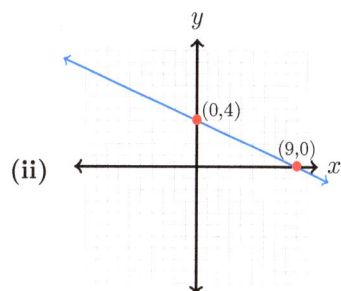

(iii)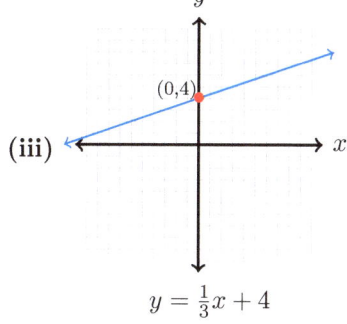

$y = \frac{1}{3}x + 4$

(iv)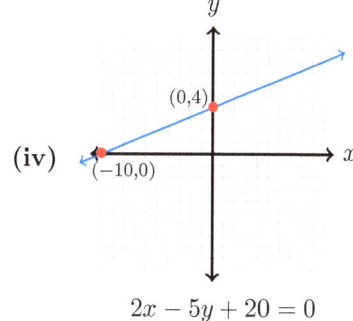

$2x - 5y + 20 = 0$

(v)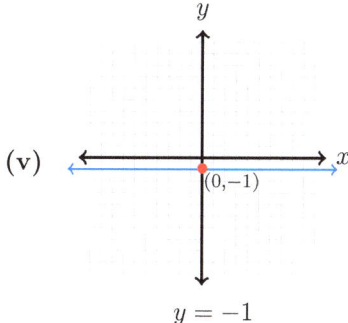

$y = -1$

(vi)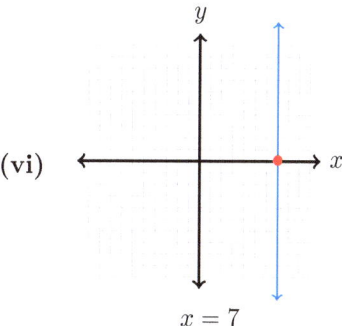

$x = 7$

Section 7.3.

Page 210.

Problems

Solve the equations for the unknowns:

1. $v - 9 = 10$
2. $y + 1.32 = 5.43$
3. $x - \frac{4}{5} = \frac{1}{3}$
4. $-35 = -\frac{5}{8}w$
5. $28 - x = 245$
6. $-7(3x - 1) + 6x = -3(x + 3)$
7. $\frac{1}{3}x - \frac{1}{5} = -\frac{4}{5}x - 1$
8. $-3(v + 3) = v - (3v + 7)$
9. $\frac{x}{2} + 6 = 2$
10. $-\frac{9}{7} + \frac{4}{3}w = -\frac{4}{3}$
11. $-\frac{3}{2}v + \frac{4}{5} = -\frac{1}{7}$
12. $4v + 2(v - 4) = -38$
13. $-5x + \frac{5}{2} = -\frac{7}{2}x - \frac{1}{5}$
14. $9w - 38 = -5(w + 2)$
15. $2(u + 5) = -4u - 8$
16. $3w + 2(w + 7) = -36$

Solve the word problems:

17. There are 196 pounds on the left plate of a scale and 250 pounds on the right plate. Suppose that you must balance the scale by removing 3-pound bags from the left plate and exactly the same number of 5-pound bags from the right plate. How many bags must you remove from each side in order to balance the scale?

18. Sung rented a truck for one day. There was a base fee of $18.95, and there was an additional charge of 74 cents for each mile driven. Sung had to pay $187.67, not including tax, when he returned the truck. For how many miles did he drive the truck?

19. A jar containing jelly beans is $\frac{1}{3}$ full. After 360 jelly beans are added, it is $\frac{5}{6}$ full. How many jelly beans can the jar hold?

20. A total of 752 tickets were sold for the school play. They were either adult tickets or student tickets. There were 52 more student tickets sold than adult tickets. How many adult tickets were sold?

Compute the value of the root in each of the following linear equations in one variable. Note that in Problems 24, 25, and 26, the equations are linear even though they may not look that way: you have to transform each of these equations a bit.

21. $3w - 4.2 = 3^2$
22. $\frac{z+3}{(2.1)^2} = 7$
23. $2t + \frac{t}{2} = t + 6$
24. $\frac{3w}{3+w} = 5$
25. $\frac{2t}{2-t} = \frac{6t}{3(2-t)} + 7$
26. $\frac{3.5(4w+6)}{2w+3} = \frac{3.5(5w-4)}{2.5w-2} + w$

SOLUTIONS

1. $v = 19$
2. $y = 4.11$
3. $x = \dfrac{17}{15}$
4. $w = 56$
5. $x = -217$
6. $x = \dfrac{4}{3}$
7. $x = -\dfrac{12}{17}$
8. $v = -2$
9. $x = -8$
10. $w = -\dfrac{1}{28}$
11. $v = \dfrac{22}{35}$
12. $v = 5$
13. $x = \dfrac{9}{5}$
14. $w = 2$
15. $u = -3$
16. $w = -10$
17. 27 bags
18. 228 miles
19. 720 jelly beans
20. 350 adult tickets
21. $3w - 4.2 = 3^2$
    $w = 4.4$
22. $\dfrac{z+3}{(2.1)^2} = 7$
    $z = 27.87$
23. $2t + \dfrac{t}{2} = t + 6$
    $t = 4$.
24. $\dfrac{3w}{3+w} = 5$
    $w = -7.5$
25. $\dfrac{2t}{2-t} = \dfrac{6t}{3(2-t)} + 7$
    the equation implies
    7=0 for any value of t
26. $\dfrac{3.5(4w+6)}{2w+3} = \dfrac{3.5(5w-4)}{2.5w-2} + w$
    $w = 0$

Section 7.4.

Page 213.

Problems

1. For each of the following systems of linear equations, either find the solution, or show that there is no solution. (That is, show that the two equations have the same slope.)

   (i) $2y = 6x + 1$,
        $3x - y = 7$;

   (ii) $y + x = 6$,
        $7y + x = 6$;

   (iii) $4y - 5x = 2$,
        $x - 5y = 8$;

   (iv) $\dfrac{2t+w+1}{t-2w} = 7$,
        $t - w = 6$;

   (v) $\dfrac{z+w-1}{z+w} = 2$,
        $z + w = -1$;

   (vi) $5y - 6x = 1$,
        $x - y = 6$.

2. The distance between two cities A and B is 385 miles. A high speed train leaves city A at noon and travels to city B at the speed of 105 miles per hour. Another train leaves city B in the direction of city A at 12:10pm. The speed of this second train is 85 miles per hour. At what time will the two trains meet, and at which distance from A will they meet? Formulate this problem as a system of two linear equations and solve it.

3. Graph the systems of linear equations and write their solutions. Note that you can also answer 'no solution' or 'infinitely many solutions'.

(i) $y = \frac{1}{3}x - 1,$
$-x + 3y = -3;$

(ii) $y = -x + 2,$
$x - 2y = -1;$

(iii) $y = -\frac{1}{4}x + 1,$
$-x - 4y = -4;$

(iv) $y = \frac{1}{3}x - 1,$
$x - 3y = -3;$

(v) $y = 2x + 3,$
$x + 3y = -5;$

(vi) $-4x - 2y = 2,$
$y = -2x - 1.$

SOLUTIONS

1. (i) $\begin{cases} 2y = 6x + 1 \\ 3x - y = 7 \end{cases} \implies \begin{cases} y = 3x + \frac{1}{2} \\ y = 3x - 7 \end{cases} \implies \begin{cases} \text{Both equations have} \\ \text{the same slope:} \\ \text{no solution.} \end{cases}$

(ii) $\begin{cases} y + x = 6 \\ 7y + x = 6 \end{cases} \implies -7x + 42 + x = 6 \implies x = 6 \text{ and } y = 0.$

(v) $\begin{cases} \frac{z+w-1}{z+w} = 2 \\ z + w = -1 \end{cases} \implies \begin{cases} \text{The two equations represent} \\ \text{the same line: } z + w = -1. \end{cases}$

(vi) $\begin{cases} 5y - 6x = 1 \\ x - y = 6 \end{cases} \implies 5y - 6y - 36 = 1 \implies y = -37 \text{ and } x = -31.$

2. Let t be the time in hours from 12pm on, and let $x = d(A, B)$, the distance in miles from A to B. So, the train from B leaves at $12\frac{1}{6}$pm. The two equations are:

$\begin{cases} x = 105t \\ 385 - x = 85(t - \frac{1}{6}) \end{cases} \implies \begin{cases} t \approx 2.10088 \\ x \approx 105 \times 2.10088 = 220.5924 \end{cases}$

The two trains will meet at approximately 2pm, 6min, and 3 sec. at approximately 221 miles from City A.

3.

(i) $\begin{cases} y = \frac{1}{3}x - 1 \\ -x + 3y = -3 \\ \text{The two equations are identical} \\ \text{Infinitely many solutions} \end{cases}$

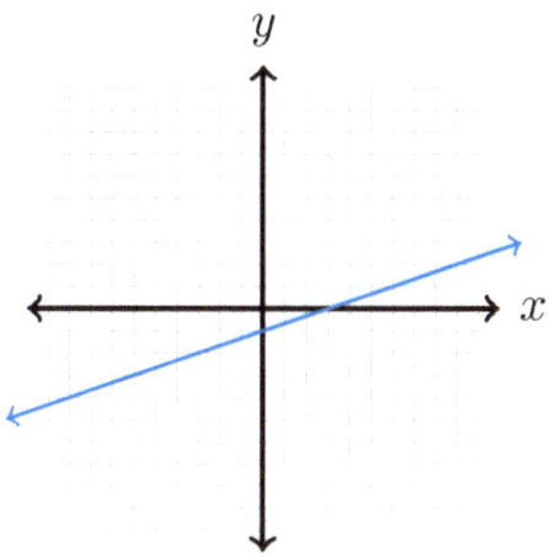

(ii) $\begin{cases} y = -x + 2 \\ x - 2y = -1 \\ \text{Solution: } (1, 1) \end{cases}$

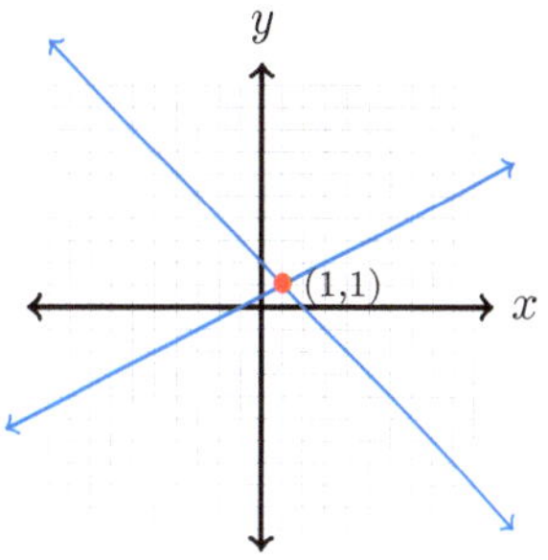

(iii) $\begin{cases} y = -\frac{1}{4}x + 1 \\ -x - 4y = -4 \\ \text{The two equations are the same} \\ \text{Infinitely many solutions} \end{cases}$

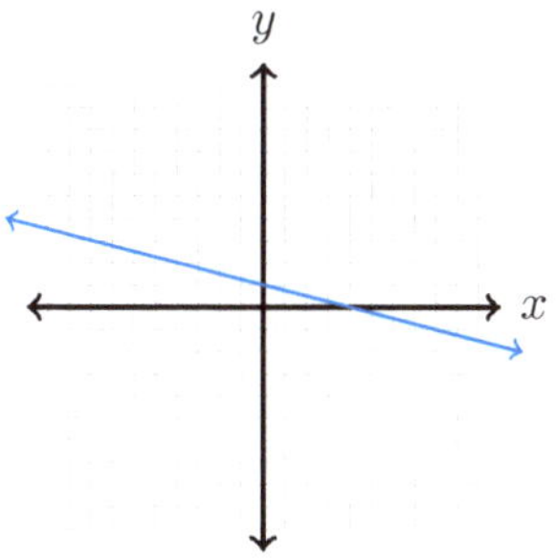

(iv) $\begin{cases} y = \frac{1}{3}x - 1 \\ x - 3y = -3 \\ \text{No solutions} \end{cases}$

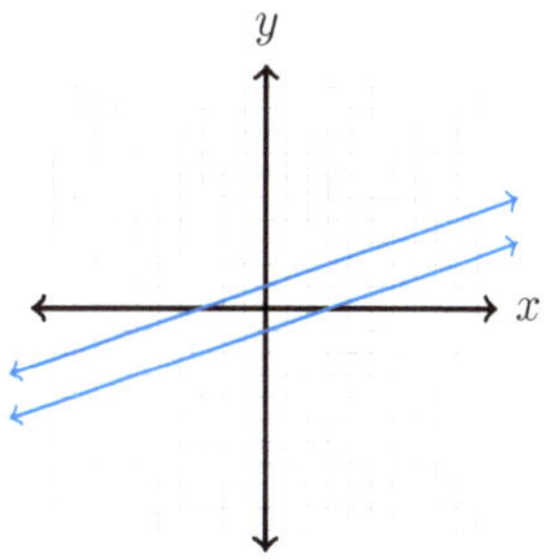

(v) $\begin{cases} y = 2x + 3 \\ x + 3y = -5 \\ \text{Solution: } (-2, -1) \end{cases}$

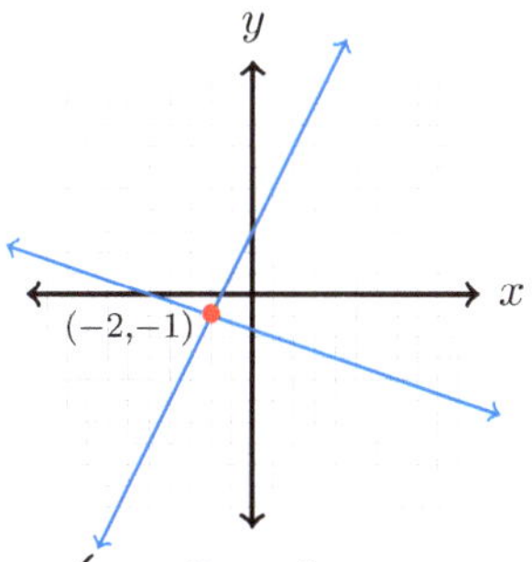

(vi) $\begin{cases} -4x - 2y = 2 \\ y = -2x - 1 \\ \text{Infinitely many solutions} \end{cases}$

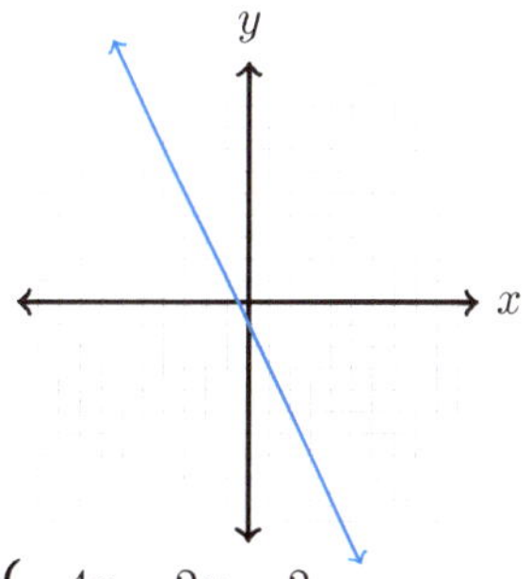

Page 215. The solutions of these two problems for the very curious students have been omitted.

Section 7.6.
Page 224.

Problems

For each of the following linear inequalities, find the interval that is the solution set. Draw a picture of the interval and write the solution set in the customary interval notation:

1. $-7 + 6v \geq -43$.
2. $-4 > 2w + 2$.
3. $2w + 1 \geq -5$.
4. $-8 < -4 + 4u$.
5. $24 > 6 + 3w$.
6. $2 - 2u \geq -22$.
7. $14 \geq -16 - 5v$.
8. $-19 \leq 4u - 7$.
9. $-\frac{1}{5}x - 3 < -5x + 5$.
10. $-4y - 7 \geq \frac{7}{5}y + 8$.

For each of the following six systems of linear inequalites, either find the interval that is the solution set, or show the solution set is empty. You must draw a picture of the intervals and write down the solution set in the customary interval notation:

1. $2x \leq 6x + 1$, $3x < 7$.
2. $3x < \frac{6}{2} + 1$, $6 \leq 7x + 1$.
3. $6z < 3.2z + 1$, $2 \leq z$.
4. $\frac{2w+1}{1-2w} \leq 7$, $w < 6$.
5. $\frac{w+1}{3w} < 2$, $w \leq -1$.
6. $3 < \frac{5}{6} - t$, $8 \leq t$.

SOLUTIONS

1. $[6, \infty)$
2. $(-\infty, -3)$
3. $[-3, \infty)$
4. $(-1, \infty)$
5. $(-\infty, 6)$
6. $(-\infty, 12]$
7. $[-6, \infty)$

8. $[-3, \infty)$

9. $(-\infty, \frac{5}{3})$

10. $(-\infty, -\frac{25}{9})$

1. $[-\frac{1}{4}, \infty) \cap (-\infty, \frac{7}{3}) = [-\frac{1}{4}, \frac{7}{3})$

2. $(-\infty, \frac{4}{3}) \cap [\frac{5}{7}, \infty) = [\frac{5}{7}, \frac{4}{3})$

4. $(-\infty, \frac{3}{8}] \cap (-\infty, 6) = (-\infty, \frac{3}{8}]$

In each of Problems 3, 5 and 6, the solutions set of each equation is indicated in red, showing that their intersection is empty.

3. $(-\infty, \frac{5}{14}) \cap [2, \infty) = \varnothing$

5. $(\frac{1}{5}, \infty) \cap (-\infty, -1] = \varnothing$

6. $(-\infty, -\frac{13}{6}) \cap [8, \infty) = \varnothing$

Page 225.

Problems for or the very curious students.

For each of the problems below, you must decide whether the union or intersection of intervals is an interval or not. If it is an interval, you must check whether it is closed, half-closed or open. Then, you have to write it down in one of the four possible forms

$$[a,b], \quad [a,b), \quad (a,b], \quad (a,b).$$

1. $(2,5) \cap [1,3]$. 2. $(1,3) \cup [0,4)$. 3. $(2,3) \cup [3,5]$.
4. $(-1,0) \cup (0,3)$. 5. $(-3,-1] \cap (2,6]$. 6. $(3,7) \cap [4,5]$.
7. $(8,9) \cup [9,10]$. 8. $(1,3) \cup [3,7)$. 9. $(2,5) \cup [2,5]$.

84   CHAPTER 7. LINEAR EQUATIONS AND LINEAR INEQUALITIES

SOLUTIONS

1. $(2,5) \cap [1,3] = [1,3]$ 2. $(1,3) \cup [0,4) = [0,4)$ 3. $(2,3) \cup [3,5] = (2,5]$

4. $(-1,0) \cup (0,3) = (-1,3)$ 5. $(-3,-1] \cap (2,6] = \varnothing$ 6. $(3,7) \cap [4,5] = [4,5]$

7. $(8,9) \cup [9,10] = (8,10]$ 8. $(1,3) \cup [3,7) = (1,7)$ 9. $(2,5) \cup [2,5] = [2,5]$

Page 229.

> **Problems for the curious students**
>
> 1. Find the solution sets of each of the six systems of linear inequalities below. Draw the solution sets on graph paper in the style of Figures 7.10 and 7.11.
>
>   (a) $2y \le x + 1$,   (b) $3x < y\frac{6}{2} + 1$,   (c) $6y < 3.2x + 1$,
> $y < x$. $6y \le 7x + 1$. $2 \le y$.
>
>   (d) $\frac{2x+1}{1+2y} \le 3$,   (e) $\frac{y-1}{x} < 2$,   (f) $3y < \frac{5}{6} - x$,
> $y < 6$. $y \le 2x + 1$. $x \le y + 1$.
>
> 2. Form a system of three linear inequalities in two variables, so that the solution set is the inside of a triangle in the coordinate plane. So, the triangle itself is not part of the solution set. Draw the solution set on a graph paper.

SOLUTIONS

1. In each of the six problems (a) to (f), each of the two inequalities corresponds to a subset of the coordinate plane which is painted in light blue and marked by **G** or **F**. The intersection is in deep blue and marked by **G**∩**F**.

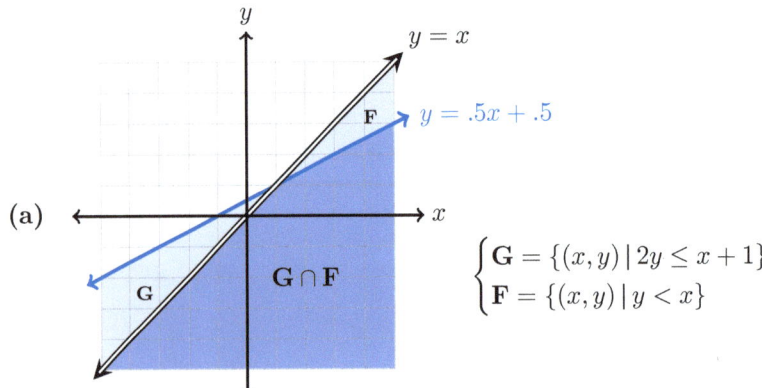

(b)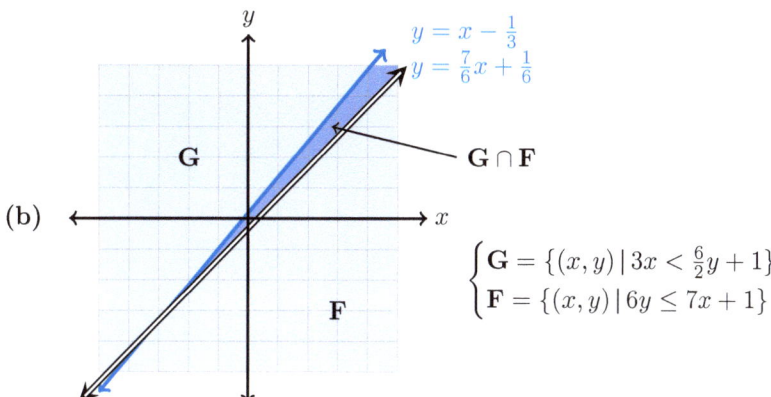

$$\begin{cases} \mathbf{G} = \{(x,y) \,|\, 3x < \frac{6}{2}y + 1\} \\ \mathbf{F} = \{(x,y) \,|\, 6y \leq 7x + 1\} \end{cases}$$

(c)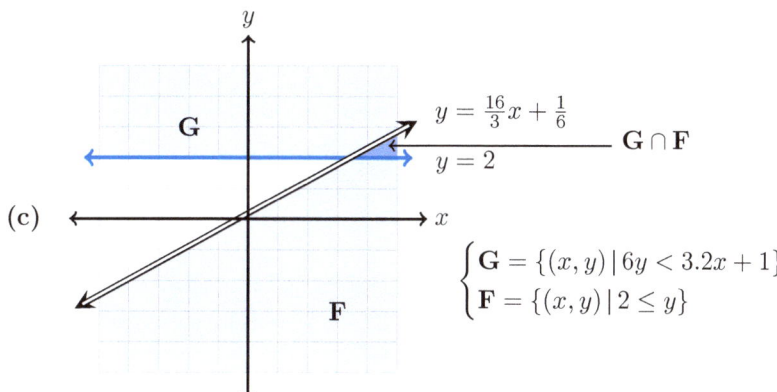

$$\begin{cases} \mathbf{G} = \{(x,y) \,|\, 6y < 3.2x + 1\} \\ \mathbf{F} = \{(x,y) \,|\, 2 \leq y\} \end{cases}$$

(d)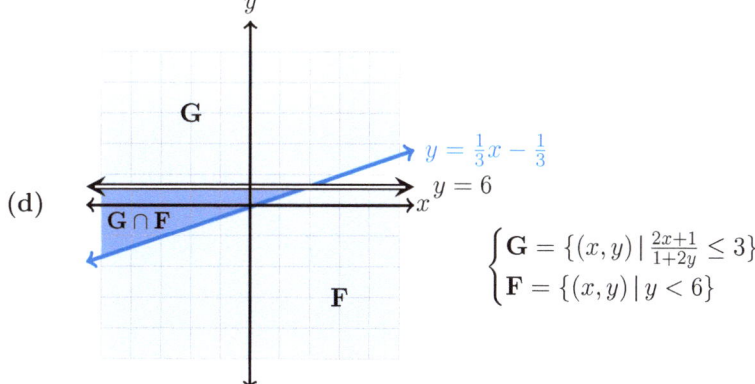

$$\begin{cases} \mathbf{G} = \{(x,y) \,|\, \frac{2x+1}{1+2y} \leq 3\} \\ \mathbf{F} = \{(x,y) \,|\, y < 6\} \end{cases}$$

(e)

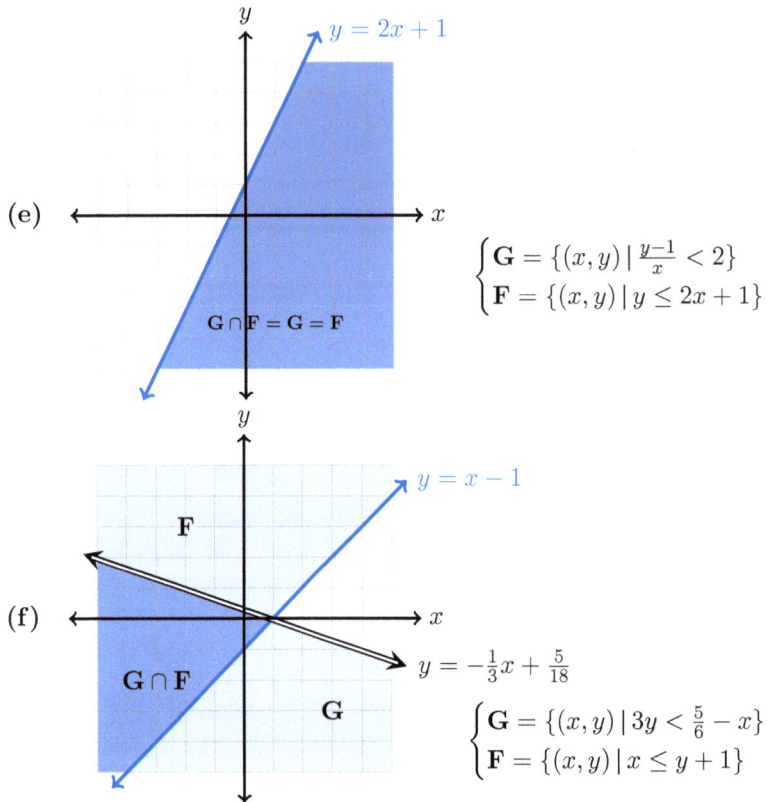

$$\begin{cases} \mathbf{G} = \{(x,y) \,|\, \frac{y-1}{x} < 2\} \\ \mathbf{F} = \{(x,y) \,|\, y \leq 2x+1\} \end{cases}$$

(f)

$$\begin{cases} \mathbf{G} = \{(x,y) \,|\, 3y < \frac{5}{6} - x\} \\ \mathbf{F} = \{(x,y) \,|\, x \leq y+1\} \end{cases}$$

2. These three inequalities define the triangle depicted below:
$$y < \frac{1}{2}x + \frac{1}{2}$$
$$y > -x - 4$$
$$y > 2x - 4.$$

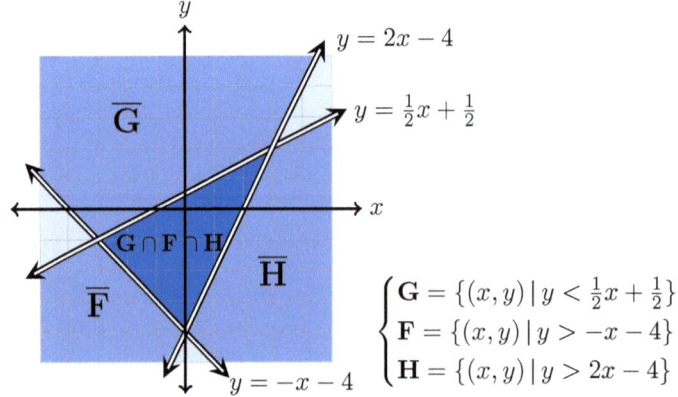

$$\begin{cases} \mathbf{G} = \{(x,y) \,|\, y < \frac{1}{2}x + \frac{1}{2}\} \\ \mathbf{F} = \{(x,y) \,|\, y > -x - 4\} \\ \mathbf{H} = \{(x,y) \,|\, y > 2x - 4\} \end{cases}$$

Chapter 8

Non-Linear Equations

Section 8.1.

Page 234.

> **Problems**
> Sketch by hand the parabolas described by the quadratic equations below. Compute the coordinates of the maximum or minimum point of the corresponding parabola and find out whether: (a) the parabola opens upward or downward; (b) the parabola touches or cuts the x-axis.
>
> 1. $y = -\frac{5}{3}x^2 - 10x - 18$.   2. $y = (x-2)^2 - 5$.   3. $y = -\frac{7}{4}x^2$.
> 4. $3y = 2x^2 - 18$.   5. $y = -.5x^2 + .05x - 1$.   6. $5y - 6x^2 = -5$.
> 7. $y = x^2 - 3x + 2$.   8. $\frac{x+y+1}{x^2} = 2$.   9. $y + x^2 = 0$.

SOLUTIONS

1. $y = -\frac{5}{3}x^2 - 10x - 18$

   (a) Downward
   (b) No touch or cut the x-axis
       Maximum: $(-3, -3)$.

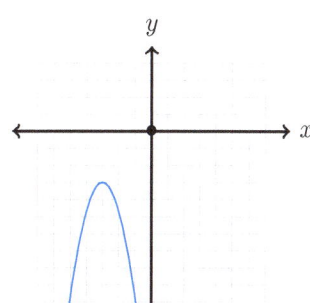

87

2. $y = (x-2)^2 - 5$

   (a) Upward

   (b) Intersects the x-axis at $(2+\sqrt{5}, 0)$ and $(2-\sqrt{5}, 0)$. Minimum: $(2, -5)$.

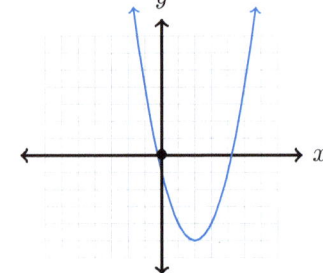

3. $y = -\frac{7}{4}x^2$

   (a) Upward.

   (b) Touches the x-axis at $(0,0)$. Minimum: $(0,0)$.

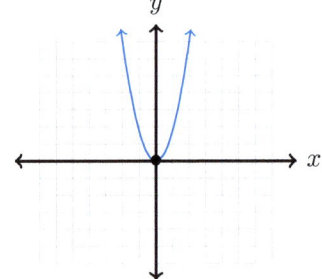

4. $3y = 2x^2 - 18$

   (a) Upward.

   (b) Intersects the x-axis at $(3,0)$ and $(-3,0)$. Minimum: $(0,-6)$.

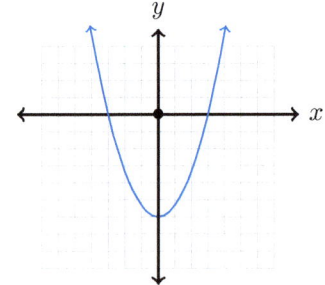

5. $y = -.5x^2 + .05x - 1$

   (a) Downward,

   (b) No touch or cut the x-axis. Maximum: $(0,-1)$.

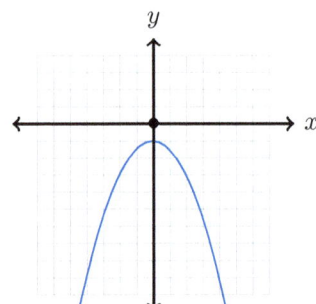

6. $5y - 6x^2 = -5$

    (a) Downward.

    (b) No touch or cut the x-axis.
        Maximum: $(0, -5)$.

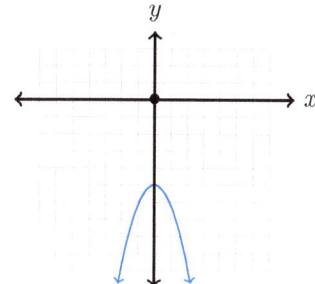

7. $y = x^2 - 3x + 2$

    (a) Upward.

    (b) Intersects the x-axis
        at $(1, 0)$ and $(2, 0)$.
        Minimum: $(1.5, -.25)$.

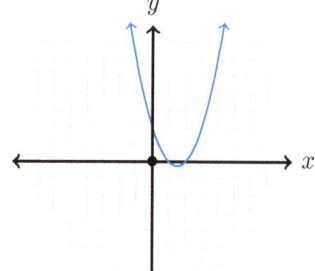

8. $\frac{x+y+1}{x^2} = 2$

    (a) Upward.

    (b) Intersects the x-axis at
        $\left(\frac{2+\sqrt{5}}{4}, 0\right)$ and $\left(\frac{2-\sqrt{5}}{4}, 0\right)$
        Minimum: $(\frac{1}{4}, -1\frac{1}{8})$.

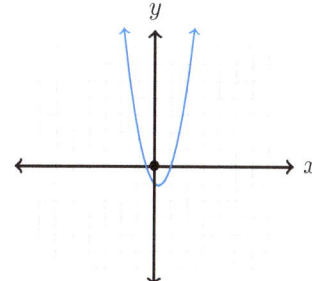

9. $y = -x^2$

    (a) Downward.

    (b) Touches the x-axis at $(0,0)$.
        Maximum: $(0, 0)$.

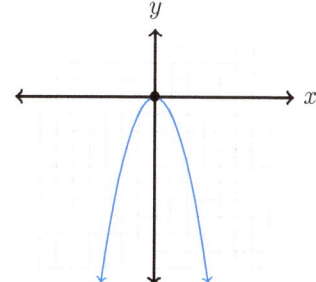

Section 8.2.

Page 236.

> **Problems**
> 1. Prove that, for any real numbers x and y, we always have
> $$(x-y)^2 = x^2 - 2xy + y^2.$$
>
> 2. Compute the squares 101^2 and 1001^2 by mental calculation. Use the Formula: $(x+y)^2 = x^2 + 2xy + y^2$.

SOLUTIONS

1. $(x-y)^2 = (x-y)(x-y)$ (by definition of the square)
$= (x-y)x - y(x-y)$ (by distributivity)
$= x^2 - yx - yx + y^2$ (by distributivity, twice)
$= x^2 - 2yx + y^2$ (because $-yx - yx = -2yx$)
$= x^2 - 2xy + y^2$ (by commutativity)

2. $(101)^2 = (100+1)^2$
$= 100^2 + 2 \times 100 + 1^2$
$= 10,000 + 200 + 1$
$= 10,201.$

$(1,001)^2 = (1,000+1)^2$
$= 1,000^2 + 2 \times 1,000 + 1^2$
$= 1,000,000 + 2,000 + 1$
$= 1,002,001.$

Section 8.3.

Page 243.

> **Problems**
>
> For each of the six equations below, compute the value of the discriminant and give the number of real solutions to the quadratic equation. Then, solve for x:
>
> 1. $2x^2 - 9x + 5 = 0$ 2. $3x^2 + 3x - 7 = 0$
> 3. $6x^2 - 6x + 1 = 0$ 4. $3x^2 + 8x + 1 = 0$
> 5. $3x^2 - 3x - 7 = 0$ 6. $4x^2 - 5x - 2 = 0$

For each of the eight equations below, compute the value of the discriminant and give the number of real solutions to the quadratic equation. (That number could be 0.) If the discriminant is non-negative, apply the completing the square method to compute the root, or the two roots, of the equation:

7. $4x^2 + 4x - 12 = 0$
8. $.5x^2 - x + .5 = 0$
9. $-9x^2 - 6x - 1 = 0$
10. $4x^2 - 5x + 5 = 0$
11. $5x^2 - 6x + 4 = 0$
12. $x^2 - \sqrt{x} + 1 = 0$
13. $5x^2 + 1.25 = 0$
14. $x^2 + x = 0$

For each of the four equations, find the value of the constant a such that the roots are those indicated on the right of the equation. As an example, for Equation 15, find the value of a such that we have
$$x^2 + x - a = (x-2)(x+3).$$

15. $x^2 + x - a = 0$ roots: 2 and -3
16. $6x^2 - 12x - a = 0$ roots: 3 and -1
17. $3x^2 - 13x + a = 0$ roots: $-\frac{2}{3}$ and 5
18. $2x^2 - 10x + a = 0$ roots: 2 and 3

SOLUTIONS

1. $2x^2 - 9x + 5 = 0$
 $b^2 - 4ac = 81 - 40 = 41$
 $41 > 0$, so 2 real solutions:
 $$x = \frac{9 \pm \sqrt{41}}{4}$$

2. $3x^2 + 3x - 7 = 0$
 $b^2 - 4ac = 9 + 84 = 93$
 $93 > 0$, so 2 real solutions:
 $$x = \frac{-3 \pm \sqrt{93}}{6}$$

3. $6x^2 - 6x + 1 = 0$
 $b^2 - 4ac = 36 - 24 = 12$
 $12 > 0$, so 2 real solutions:
 $$x = \frac{6 \pm \sqrt{12}}{12}$$

4. $3x^2 + 8x + 1 = 0$
 $b^2 - 4ac = 64 - 12 = 52$
 $52 > 0$, so 2 real solutions:
 $$x = \frac{-8 \pm \sqrt{52}}{6}$$

5. $3x^2 - 3x - 7 = 0$
 $b^2 - 4ac = 9 + 84 = 93$
 $93 > 0$, so 2 real solutions:
 $$x = \frac{3 \pm \sqrt{93}}{6}$$

6. $4x^2 - 5x - 2 = 0$
 $b^2 - 4ac = 25 + 32 = 57$
 $57 > 0$, so 2 real solutions:
 $$x = \frac{5 \pm \sqrt{57}}{8}$$

7. $4x^2 + 4x - 12 = 0$
   $b^2 - 4ac = 16 + 192 = 208$
   $208 > 0$, so 2 real solutions:
   $$x = \frac{-4 \pm \sqrt{208}}{8} = \frac{-1 \pm \sqrt{13}}{2}$$

8. $.5x^2 - 1x + .5 = 0$
   $b^2 - 4ac = 1 - 4 \times .25 = 0$
   so 1 real solution: $x = 1$

9. $-9x^2 - 6x - 1 = 0$
   $b^2 - 4ac = 36 - 36 = 0$
   so 1 real solution: $x = -\dfrac{1}{3}$

10. $4x^2 - 5x + 5 = 0$
    $b^2 - 4ac = 25 - 80 = -55$
    $-55 < 0$, so no real solution.

11. $5x^2 - 6x + 4 = 0$

    $b^2 - 4ac = 36 - 80 = -44$
    $-44 < 0$, so 0 real solutions.

12. $x^2 - \sqrt{x} + 1 = 0$

    $b^2 - 4ac = 1 - 4 \times 1 \times 1 = -3$
    $-3 < 0$, so no real solution.

The next two problems need no computation.

13. $5x^2 + 1.25 = 0$ this implies
    $x^2 = -\dfrac{1.25}{5} = -.25 < 0$
    so no real solution.

14. $x^2 + x = 0 = x(x+1)$
    so 2 real solutions: $x = 0, -1$

15. $x^2 + x - a$
    $= (x - 2)(x + 3)$
    $= x^2 + x - 6$
    $a = 6$

16. $6x^2 - 12x - a$
    $= (6x - 18)(x + 1)$
    $= 6x^2 - 12x - 18$
    $a = 18$

17. $3x^2 - 13x + a$
    $= (3x + 2)(x - 5)$
    $= 3x^2 - 13x + 10$
    $a = 10$

18. $2x^2 - 10x + a$
    $= (2x - 4)(x - 3)$
    $= 2x^2 - 10x + 12$
    $a = 12$

Page 246.

> **Problems for the very curious students**
>
> 1. Show that the coordinates of the focus F of a parabola of equation $y = a^2 + bx + c$ are
>
> $$F = \left(\tfrac{-b}{2a}, c - \tfrac{b^2-1}{4a}\right).$$
>
> These coordinates are (s,t). You should use the three equations
>
> $$a = \frac{1}{2(t-k)}, \quad b = -\frac{s}{t-k} \quad \text{and} \quad c = \frac{s^2 + t^2 - k^2}{2(t-k)}, \quad (8.1)$$
>
> consistent with $y = ax^2 + bx + c$ to get the formulas for s and t in terms of a, b and c. Start by observing that $t - k = \frac{1}{2a}$. (Hint: the basic formula $t^2 - k^2 = (t-k)(t+k)$ may be useful.)
>
> 2. Using the same method, prove that the equation of the directrix is $y = c - \frac{b^2+1}{4a}$.
>
> 3. Compute the coordinates and the equation of the directrix for the three equations describing parabolas in Problems 1-3 on page 87.
>
> 4. It is not possible for a directrix to cut or even touch its parabola. Can you explain why?

SOLUTIONS

1. We have to solve, for s and t, the three equations in (8.1). Using these three equations in succession, we get

$$t - k = \frac{1}{2a} \quad \text{(from the first equation in (8.1))}$$

$$s = -b(t-k) = -\frac{b}{2a} \quad \left(\begin{array}{c}\text{from the second equation in (8.1))} \\ \text{and the equation above}\end{array}\right)$$

$$c = \frac{\frac{b^2}{4a^2} + (t-k)(t+k)}{2(t-k)} \quad \left(\begin{array}{c}\text{from the third equation in (8.1)} \\ \text{and the two equations above}\end{array}\right)$$

$$= \frac{\frac{b^2}{4a^2}}{\frac{1}{a}} + \frac{t+k}{2} \quad \left(\begin{array}{c}\text{by simplification and the value} \\ \text{of } t-k \text{ in the top equation}\end{array}\right)$$

$$= \frac{b^2}{4a} + \frac{t}{2} + \frac{t}{2} + \frac{1}{4a} \quad \text{(because } k = t - \frac{1}{2a}\text{)}$$

$$= \frac{b^2 - 1}{4a} + t \quad \text{which yields} \quad t = c - \frac{b^2-1}{4a}.$$

This implies: $\quad F = (s,t) = \left(\tfrac{-b}{2a}, c - \tfrac{b^2-1}{4a}\right).$

2 and 3. We omit the solutions to these two problems.

4. We proceed by contradiction. Suppose that there is a point P that is common to the parabola and to its directrix ℓ. Since by definition of the directrix, the distance of P to ℓ must be equal to its distance to the focus F of the parabola, we must have $d(P, F) = 0$, and so $F = P$, that is, P is the focus of the parabola. Take any other point P' in the parabola that is not a point of ℓ, and let $Q \in \ell$ be such that $\overleftrightarrow{P'Q}$ is perpendicular to ℓ. The distance $P'Q$ of P' to ℓ must be equal to the distance PP' of P' to the focus P. But $\triangle P'QP$ is a right triangle, with hypothenuse $\overline{PP'}$. So, the hypothese that $PP' = P'Q$ is in contradiction to the Pythagorean Theorem. (Note that this argument is valid if we suppose that ℓ contains two points of the parabola.)

Section 8.5.

Page 251.

> **Problems**
>
> In the problems below, you must decide whether the expression is a monomial, a binomial or a trinomial or some other polynomial. You must also find its degree.
>
> 1. $xyxw$.
> 2. $x^2t^2y^3 + x^7 + 6x - t^5$.
> 3. $x^5 + 4xyz + t^4$.
> 4. $3y + 2x^2 - 18$.
> 5. $y^3x^2 - .5x^2 + x - 1$.
> 6. $5y - 6x^2 - 5$.
> 7. $y^6 - x^2 - 3x + 2$.
> 8. x^3y^6.
> 9. $x^3y^3 + 2z^4$.

SOLUTIONS

1. $xyxy = x^2yw$
   monomial
   degree: 4

2. $x^2t^2y^3 + x^7 + 6x - t^5$
   polynomial
   degree: 7

3. $x^5 + 4xyz + t^4$
   trinomial
   degree: 5

4. $3y + 2x^2 - 18$
   trinomial
   degree: 2

5. $y^3x^2 - .5x^2 + x - 1$
   polynomial
   degree: 5

6. $5y - 6x^2 - 5$
   trinomial
   degree: 2

7. $y^6 - x^2 - 3x + 2$
   polynomial
   degree: 6

8. x^3y^6
   monomial
   degree: 9

9. $x^3y^3 + 2z^4$
   binomial
   degree: 6

Page 253.

> **Problems**
>
> Each of the problems asks for the addition or subtraction of two or more monomials. In some cases, the monomials are like, but not always. Simplify the result if you can. First, check whether the monomials are like. You may have to use the commutativity property to do that.
>
> 1. $3x^2tz - 6tzx^2 - tx^2z$.
> 2. $6ztw^3 - 3tw^3z$.
> 3. $2z^3tz + 5z^2tz^2$.
> 4. $6ztz^2x - 6z^2tx^2z$.
> 5. $xy^2z^3 - 2z^3xy^2 + y^2z^3x$
> 6. $ztx^2x - 2x^2z^2tz$.

SOLUTIONS

1. $3x^2tz - 6tzx^2 - tx^2z$
   $= -4x^2tz$

2. $6ztw^3 - 3tw^3z$
   $= 3ztw^3$

3. $2z^3tz + 5z^2tz^2$
   $= 7z^4t$

4. $6ztz^2x - 6z^2tx^2z$
   $= 6z^3tx - 6z^3x^2t$

5. $xy^2z^3 - 2z^3xy^2 + y^2z^3x$
   $= z^3y^2x - 2z^3y^2x + z^3y^2x$
   $= 0$

6. $ztx^2x - 2x^2z^2tz$
   $= x^3zt - 2z^3x^2t$

Page 254.

> **Problems**
>
> In each problem, carry out the multiplication and simplify the result as much as possible.
>
> 1. $3w^2tz \times .2tzw^3 \times tx^2z$.
> 2. $6ztw^3 \times 3tw^3z$.
> 3. $2z^3tz \times 5z^2tz^2$.
> 4. $-6ztz^2x \times 6z^2tx^2z$.
> 5. $z^3wz \times 3z^2tz^4$.
> 6. $2zx^2tz^2x \times (-3z^2tx^2z)$.
> 7. $(2xt + z^2y)(xy + z^2t)$.
> 8. $(x + z)(tzx + tx^2 + xz)$.
> 9. $(2x^2y - z^2t)(tz^2 - yx^2)$.
> 10. $(x + y - w)(x - y + w)$.

SOLUTIONS

1. $3w^2tz \times .2tzw^3 \times tx^2z$
$= .6w^5z^3t^3x^2$

2. $6ztw^3 \times 3tw^3z$
$= 18w^6z^2t^2$

3. $2z^3tz \times 5z^2tz^2$
$= 10z^8t^2$

4. $-6ztz^2x \times 6z^2tx^2z$
$= -36z^6x^3t^2$

5. $z^3wz \times 3z^2tz^4$
$= 3z^{10}wt$

6. $2zx^2tz^2x \times (-3z^2tx^2z)$
$= -6z^6x^5t^2$

7. $(2xt + z^2y)(xy + z^2t)$
$= 2x^2ty + 2z^2t^2x$
$+ z^2y^2x + z^4ty$

8. $(x+z)(tzx + tx^2 + xz)$
$= 2x^2tz + x^3t + x^2z + z^2tx + z^2x$

9. $(2x^2y - z^2t)(tz^2 - yx^2)$
$= 3z^2x^2ty - z^4t^2 - 2x^4y^2$

10. $(x+y-w)(x-y+w)$
$= x^2 - y^2 - w^2 + 2wy$

Page 255.

Problem

In each of the two problems below, give the list of the common divisors of the three monomials.
1. z^2xy^3, z^3xy, $xzwy^2$.
2. $4w^3yz^2$, $2wxy$, $3wy^2$.

SOLUTIONS

Monomials	Common divisors
1. z^2xy^3, z^3xy, $xzwy^2$	x, y, z, xy, xz, yz, xyz, and 1
2. $4w^3yz^2$, $2wxy$, $3wy^2$	y, w, yw, and 1

Page 256.

Problems for the very curious students

In each of the three problems below, give the greatest common divisor of the three monomials.
1. z^2xy^3, z^3xy, $xzwy^2$.
2. $4w^3yz^2$, $2wxy$, $3wy^2$.
3. $18xy$, $6y^2wx$, $3xy$.

SOLUTIONS

	Monomials	GCD
1.	z^2xy^3, z^3xy, $xzwy^2$	xyz
2.	$4w^3yz^2$, $2wxy$, $3wy^2$	yw
3.	$18xy$, $2wxy$, $3wy^2$	y

Review Problems.

Page 258.

Review Problems

Simplify:

1. $(-3u^2 - 7u + 8) - (-u^2 - 4u + 8) + (2u^2 + 5u + 1)$
2. $(-5x^2 + 5x - 9) + (-8x^2 - 3x - 5) - (-3x^2 + 2x + 4)$
3. $(9y^2 + 5y - 6) - (-4y^2 + 7y + 7) + (-7y^2 + 6y + 9)$
4. $(6x^2 + 9x + 9) + (-5x^2 + 8x + 5) - (4x^2 + 6x - 5)$

Multiply, then simplify your answer:

5. $(x-8)(x+1)(u+1)(u+3)$ 6. $(u+2y)(u-6y)$
7. $(6x+y-5)(7x-6y+4)$ 8. $(v-4w)(7v-6w-7)$

Rewrite without parentheses and simplify:

$9(5v-2)^2$ $10(5x+6)^2$ $11(3u-4)^2$ $12(3w+7)^2$

Find the GCD in each case:

13. $24w^8v^4y^3$ and $28w^4y^3$
14. $11v^3w^5$ and $22v^7w^2x^8$
15. $20x^2w^8$ and $8x^7wv^6$
16. $10y^8x^2$ and $16y^8x^6v^5$
17. $3t^2y^2$, xt, $3ty$
18. 49, 7, $7x$
19. x^2yw^4, $3yw^2$, $6y^2w^6$, xyw^2
20. 7, 49, , $35x$, $7x$

Divide each of the polynomials below by the GCD of its terms and simplify the result as much as possible:

21. $3t^2y^2 + wty + 3ty^2$

22. $9x^2t + 12xt + 3x^3yt$

23. $x^2yw^4 + 3yw^2 + 6y^2w^6 + xyw^2$

24. $7w^2x + 49xw + 35x^3w^3 + 7x^2w^2$

Give the degree of the polynomial:

25. $y^6v - 3w^8 - 4 + 16v^5w^2y^3$ 26. $-w^9v^5 - 6 + 3v^3y^8w^2 + 6y$

27. $-2x^2 - 7 + 16ux^4w^2 - w^3u^3$ 28. $-vy^6 + 5 - 5w^9 + 2y^4w^2v^2$

SOLUTIONS

1. $(-3u^2 - 7u + 8) - (-u^2 - 4u + 8) + (2u^2 + 5u + 1) = 2u + 1$

2. $(-5x^2 + 5x - 9) + (-8x^2 - 3x - 5) - (-3x^2 + 2x + 4) = -10x^2 - 18$

3. $(9y^2 + 5y - 6) - (-4y^2 + 7y + 7) + (-7y^2 + 6y + 9) = 6y^2 + 4y - 4$

4. $(6x^2 + 9x + 9) + (-5x^2 + 8x + 5) - (4x^2 + 6x - 5) = -3x^2 + 11x + 19$

5. $(x - 8)(x + 1)(u + 1)(u + 3) = (x^2 - 7x - 8)(u^2 + 4u + 3)$
   $= x^2u^2 + 4x^2u + 3x^2 - 7u^2x - 28xu - 21x - 8u^2 - 32u - 24$

6. $(u + 1)(u + 3) = u^2 + 4u + 3$

7. $(6x + y - 5)(7x - 6y + 4) = 42x^2 - 6y^2 - 29xy - 11x + 34y - 20$

8. $(v - 4w)(7v - 6w - 7) = 7v^2 + 24w^2 - 34vw - 7v + 28w$

9. $(5v - 2)^2 = 25v^2 - 20v + 4$

10. $(5x + 6)^2 = 25x^2 + 60 + 36$

11. $(3u - 4)^2 = 9u^2 - 24u + 16$

12. $(3w + 7)^2 = 9w^2 + 42w + 49$

13. $24w^8v^4y^3$, $28w^4y^3$
    GCD: $4w^4y^3$

14. $11v^3w^5$, $22v^7w^2x^8$
    GCD: $11v^3w^2$

15. $20x^2w^8$, $8x^7wv^6$
    GCD: $4x^2w$

16. $10y^8x^2$, $16y^8x^6v^5$
    GCD: $2y^8x^2$

17. $3t^2y^2$, xt, $3ty$
    GCD: t

18. 49, 7, $7x$
    GCD: 7

19. x^2yw^4, $3yw^2$, $6y^2w^6$, xyw^2
    GCD: yw^2

20. 7, 49, , $35x$, $7x$
    GCD: 7

21. $3t^2y^2 + wty + 3ty^2$
    $= ty(3ty + w + 3y)$

22. $9x^2t + 12xt + 3x^3yt$
    $= 3xt(3x + 4 + x^2y)$

23. $x^2yw^4 + 3yw^2 + 6y^2w^6 + xyw^2$
    $= yw^2(x^2w^2 + 3 + 6yw^4 + x)$

24. $7w^2x + 49xw + 35x^3w^3 + 7x^2w^2$
    $= 7xw(w + 7 + 5x^2w^2 + xw)$

25. $y^6v - 3w^8 - 4 + 16v^5w^2y^3$
    degree: 10

26. $-w^9v^5 - 6 + 3v^3y^8w^2 + 6y$
    degree: 14

27. $-2x^2 - 7 + 16ux^4w^2 - w^3u^3$
    degree: 7

28. $-vy^6 + 5 - 5w^9 + 2y^4w^2v^2$
    degree: 9

Chapter 9

Lines and Angles

Section 9.1.

Page 262.

> **Problems**
>
> For each of the problems below, give the name of the object pictured in red in Problems 1, 2 and 4, and in black in Problem 3, and indicate how it should be labeled, using the conventions for the notation of a segments, a ray and a line containing the points P and Q: $\overline{PQ}$ is a segment, $\overrightarrow{PQ}$ is a ray, and $\overleftrightarrow{PQ}$ is a line.
>
> 1. ←•—•→ A B
> 2. ←•—•→ A B
> 3. ←•—•→ A B
> 4. ←•—•→ A B

SOLUTIONS

    1. segment; $\overline{AB}$ 2. ray; $\overrightarrow{AB}$ 3. line; $\overleftrightarrow{AB}$ 4. ray; $\overrightarrow{BA}$

Page 265. We skip the solution of this problem.

Section 9.2.

Page 271.

Problems

Solve the problems:

1. An angle measures 138 degrees. What is the measure of its supplement?

2. An angle measures 124 degrees. What is the measure of its supplement?

3. An angle measures 54 degrees. What is the measure of its complement?

4. An angle measures 35 degrees. What is the measure of its complement?

Classify the four angles of the quadrilaterals as acute, obtuse, or right:

5. Quadrilateral ABCD with angles: A—90°, B—105°, C—75°, D—90°.

6. Quadrilateral ABCD with angles: A—90°, B—60°, C—140°, D—70°.

SOLUTIONS

1. 42 degrees   2. 56 degrees   3. 36 degrees   4. 55 degrees

5. A—right, B—obtuse, C—acute, D—right,

6. A—right, B—acute, C—obtuse, D—acute.

Section 9.3.

Page 273.

Problems

1. How many non-reflex angles (that is: angles measuring less than $180°$) can be formed with 4 rays with a common vertex?

2. (Continuation.) What is the smallest and the greatest number of pairs of non-reflex adjacent angles?

> 3. **For the very curious students.** How many non-reflex angles, at most, can be formed with n rays with a common vertex? The number n is any positive integer greater than 2. What is the smallest and the greatest number of pairs of adjacent angles?

SOLUTIONS

1. Six non-reflex angles can be formed. In the figure below on the left we have:
$$\angle AYX, \angle XYZ, \angle ZYW, \angle AYZ, \angle XYW, \angle AYW.$$

2. The smallest and the largest number of pairs of non-reflex adjacent angles is the same, namely 4 pairs. Using the figure on the left, we have:
$$(\angle AYX, \angle XYZ), (\angle XYZ, \angle ZYW),$$
$$(\angle AYZ, \angle ZYW), (\angle AYX, \angle XYW).$$

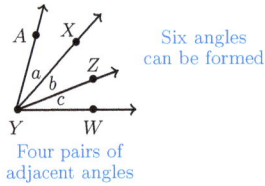
Six angles can be formed

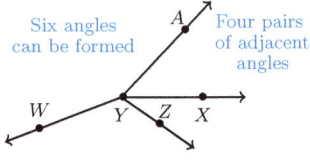
Six angles can be formed Four pairs of adjacent angles

Four pairs of adjacent angles

3. The number of non-reflex angles that can be formed with n rays with a common vertex is
$$\tfrac{n(n+1)}{2} = n + n - 1 + \ldots + 2 + 1.$$

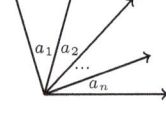

This is a well known result: see Volume 1.
The situation is illustrated by the figure on the right and by the table below.

	Cases	Number
Single angle	$\angle a_1, \angle a_2, \ldots, \angle a_n$	n
Summing two adjacent angles	$\angle a_1 + a_2, \angle a_2 + a_3, \ldots, \angle a_{n-1} + a_n$	$n-1$
...	...	...
Summing $n-1$ angles	$\angle a_1 + a_2 + \ldots + a_{n-1}, \angle a_2 + a_3 + \ldots, +a_n$	2
Summing n angles	$a_1 + a_2 + \ldots + a_n$	1
Total:		$\tfrac{n(n+1)}{2}$

Page 278.

Problems

1. Give the types of each of the angles in the figure of Problem 6 below.

2-3. We omit the solution of these two problems.

4. Prove that, if one of the eight angles produced by the intersection of two parallel lines with a transversal is a right angle, then all the other angles are also right angles. Be careful how you do that. There is no need to give eight proofs, each starting with a sentence: "Suppose that $\angle n$ is a right angle" (with $n = 1, 2 \ldots 8$). However, you have to give a good reason for not doing that.

5. We omit this proof.

Give a pair of corresponding angles, a pair of alternate exterior angles, and a pair of alternate interior angles for each of the figures below:

6.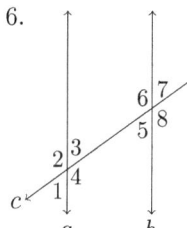

7.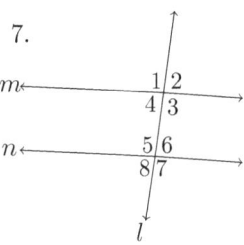

Give one pair of supplementary angles and one pair of vertical angles for each of the figures below:

8.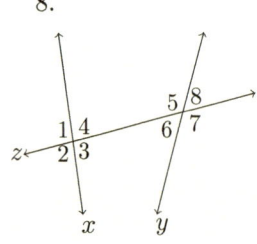

9.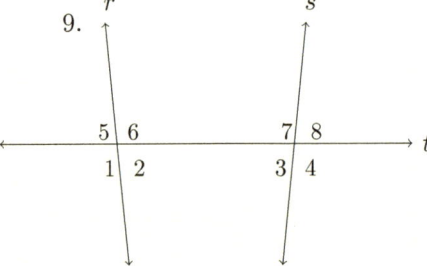

Find the values of x and z in each of the figures of Problems 10-13 below:

10.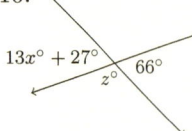

11.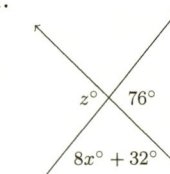

12.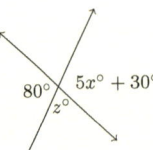

13.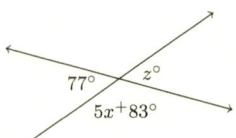

In each of the four figures below, we have $g \| k$. (That is, the lines g and k are parallel.) Find the values of x and z.

14.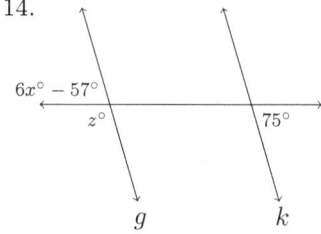

15.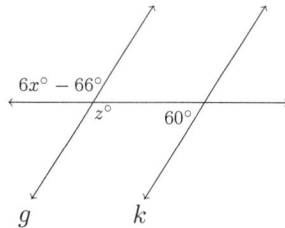

16.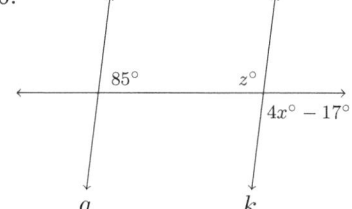

17.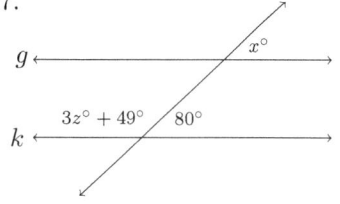

SOLUTIONS

1. Types of angles in the figure of Problem 6:

Interior angles:	$\angle 3$, $\angle 4$, $\angle 5$, $\angle 6$
Exterior angles:	$\angle 1$, $\angle 2$, $\angle 7$, $\angle 8$
Alternate interior angles:	$\angle 4$ and $\angle 6$, $\angle 5$ and $\angle 3$
Alternate exterior angles:	$\angle 1$ and $\angle 7$, $\angle 2$ and $\angle 8$
Same side interior angles:	$\angle 4$ and $\angle 5$, $\angle 3$ and $\angle 6$
Same side exterior angles:	$\angle 1$ and $\angle 8$, $\angle 2$ and $\angle 7$
Corresponding angles:	$\angle 1$ and $\angle 5$, $\angle 2$ and $\angle 6$
	$\angle 4$ and $\angle 8$, $\angle 3$ and $\angle 7$

2-3. Omitted.

4. Proof that if one of the eight angles produced by the intersection of two parallel lines with a transversal is a right angle, then all the other angles are also right angles:

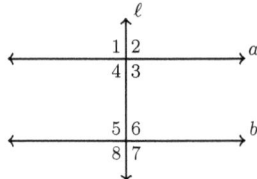

We recall Postulate [L4]:

[L4] *The corresponding angles produced by a transversal intersecting two parallel lines are congruent.*

PROOF.

(1) line ℓ and line a are perpendicular (definition of right angle)

(2) $m\angle 1 = m\angle 2 = m\angle 3 = m\angle 4 = 90°$ (definition of perpendicular)

(3) $m\angle 5 = m\angle 6 = m\angle 7 = m\angle 8 = 90°$ (Postulate [L4]).

(4) $m\angle i = 90°$ for $1 \leq i \leq 8$ (by Line (2) and (3))

□

5. This proof relies on Theorem 9.C in Volume 1. We leave the details to the student.

6. Angle pairs:
   A pair of corresponding angles: $\angle 1$ and $\angle 5$
   A pair of alternate exterior angles: $\angle 1$ and $\angle 7$
   A pair of alternate interior angles: $\angle 3$ and $\angle 5$

7. Angle pairs:
   A pair of corresponding angles: $\angle 1$ and $\angle 5$
   A pair of alternate exterior angles: $\angle 1$ and $\angle 7$
   A pair of alternate interior angles: $\angle 3$ and $\angle 5$

8. Angle pairs:
   A pair of supplementary angles: $\angle 1$ and $\angle 2$
   A pair of vertical angles: $\angle 1$ and $\angle 3$

9. Angle pairs:
   A pair of supplementary angles: $\angle 1$ and $\angle 2$
   A pair of vertical angles: $\angle 1$ and $\angle 6$

Page 279.

10.	$x = 3$	11.	$x = 9$	12.	$x = 10$	13.	$x = 4$
	$z = 114$		$z = 76$		$z = 100$		$z = 77$
14.	$x = 22$	15.	$x = 31$	16.	$x = 28$	17.	$x = 80$
	$z = 105$		$z = 120$		$z = 95$		$z = 17$

Page 283. We omit the solutions of these problems.

Chapter 10

Triangles and Quadrangles

Section 10.1.

Page 287.

We recall:

THEOREM 10.A. *The sum of the measures of the angles of a triangle is equal to* $180°$.

Problems
1. Prove Corollary 10.A2.
2. Prove Corollary 10.A3.
3. Prove Corollary 10.A4.
4. Prove Corollary 10.A5.

SOLUTIONS

1. COROLLARY 10.A2: *Each angle of an equilateral triangle measures* $60°$.

   PROOF. From Theorem 10.A, we know that the sum of the measures of the angles of a triangle is equal to $180°$. By definition, any equilateral triangle is equiangular. So, each of the angles of an equilateral triangle measures $\frac{180°}{3} = 60°$. □

2. COROLLARY 10.A3: *In a triangle, if one angle is obtuse or right, the two other angles must be acute.*

   PROOF. Let $\angle a$, $\angle b$ and $\angle c$ be the three angles of a triangle, and suppose that $m\angle c \geq 90°$. By Theorem 10.A, we have $m\angle a + m\angle b + m\angle c = 180°$. This implies
   $$m\angle a + m\angle b = 180° - m\angle c \leq 180° - 90° = 90°$$
   which yields
   $$m\angle a + m\angle b \leq 90°.$$
   Since $\angle a, \angle b > 0$, we get $m\angle a < 90°$ and $m\angle b < 90°$. □

3. COROLLARY 10.A4: *If two angles of one triangle are congruent to two angles of some other triangle, then the third angles of each triangle are also congruent.*

   PROOF. Let $\angle a_1, \angle a_2$ and $\angle a_3$ be the three angles of the first triangle, and let $\angle b_1, \angle_2$ and $\angle b_3$ be the three angles of the second triangle, and suppose that $m\angle a_1 = m\angle b_1$ and $m\angle a_2 = m\angle b_2$. This implies
   $$m\angle a_1 + m\angle a_2 = m\angle b_1 + m\angle b_2. \qquad (\star)$$
   By Theorem 10.A, we have
   $$m\angle a_1 + m\angle a_2 + m\angle a_3 = m\angle b_1 + m\angle b_2 + m\angle b_3 = 180°. \qquad (\star\star)$$
   Subtracting each side of $(\star)$ from the corresponding side of the first equation in $(\star\star)$, we obtain $m\angle a_3 = m\angle b_3$. □

4. COROLLARY 10.A5: *The measure of an exterior angle $\angle a$ of a triangle is equal to the sum of the two interior angles not adjacent to $\angle a$.*

   PROOF. We write $\angle b$ for the angle of the triangle adjacent to $\angle a$, and $\angle c$ and $\angle d$ for the two other angles. By Theorem 10.A, and using the fact that the measure of an angle made of two collinear rays is equal to 180°, the situation of the corollary is described by the equations
   $$180° = \angle b + \angle c + \angle d = \angle a + \angle b,$$
   which yields: $\angle c + \angle d = \angle a$. □

Page 296-297.

Problems

For each of the eight triangles 1 to 4 below and 5 to 8 on the next page, state whether it is acute, obtuse, right, isosceles, scalene or equilateral. More than one of these adjectives may apply in a particular case.

1.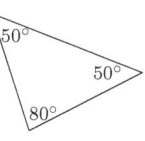

2.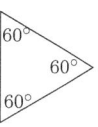

3.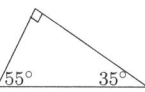

4.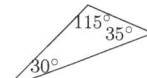

CONTINUATION. Which of the triangles 5 to 8 is acute, obtuse, right, isosceles, scalene or equilateral?

5.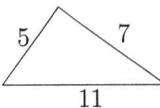

6.

7.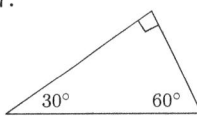

8.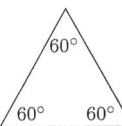

In each of Problems 9 to 14, find the measure of the angle x.

In Problems 15 and 16, find the length of the segment x, rounding your answer to the nearest tenth if necessary.

9.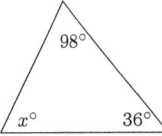

10.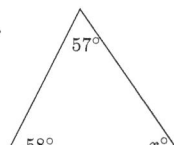

11.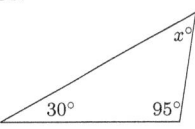

12.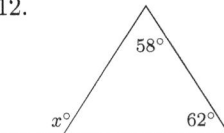

13.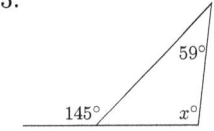

14.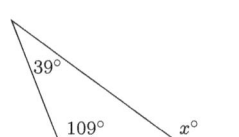

15.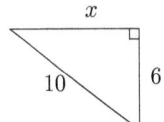

16.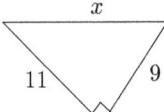

SOLUTIONS

In the table below, each line concerns one of the 8 triangles in Problems 1-8, and each column from 2-7 corresponds to one of the 6 properties. An x in a column indicates that the triangle of that line has the property.

Problems	Acute	Isoceles	Equilateral	Right	Obtuse	Scalene
1.	x	x				
2.	x	x	x			
3.				x		
4.						x
5.					x	x
6.	x	x				
7.				x		
8.	x	x	x			

Each of the columns of the table below contains the response to one of Problems 9-16. The value of x^o is indicated below the problem number.

Problems							
9.	10.	11.	12.	13.	14.	15.	16.
46°	65°	55°	120°	86°	148°	8	≈ 14.2

Section 10.3.

Page 309-310.

> **Problems**
>
> Find the area of each of the figures below. Include correct units in your answer:
>
>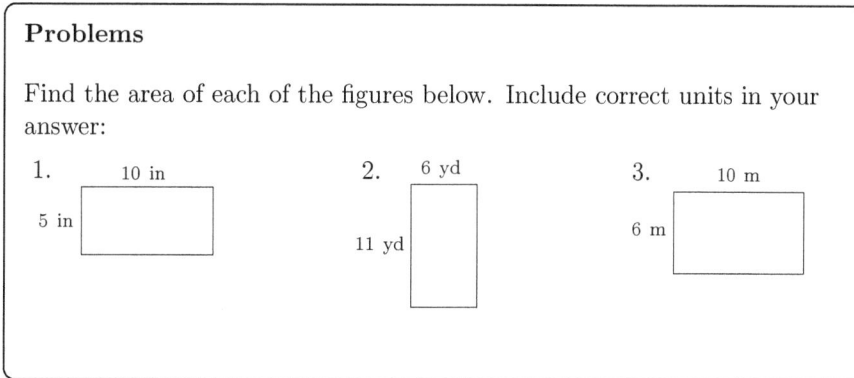

CHAPTER 10. TRIANGLES AND QUADRANGLES

Continuation

4.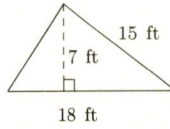
5.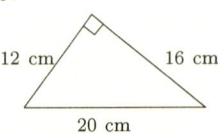
6.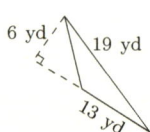
7.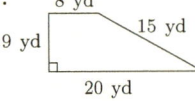
8.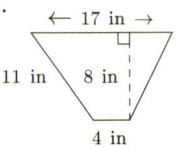
9.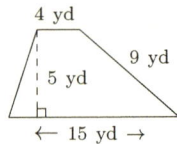

Solve the word problems:

10. The area of a rectangular field is 4526 square yards. If the length of the field is 73 yards, what is its width?

11. A rectangle is removed from an isosceles right triangle to create the shaded region shown on the right. Find the area of this shaded region. Be sure to include correct units in your answer.

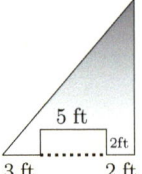

12. A rectangle is removed from an isosceles right triangle to create the shaded region shown on the right. Find the area of this shaded region. Be sure to include correct units in your answer.

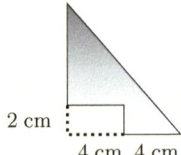

SOLUTIONS

Problems					
1.	2.	3.	4.	5.	6.
$50\,in^2$	$66\,ya^2$	$60\,m^2$	$63\,ft^2$	$96\,cm^2$	$39\,ya^2$
7.	8.	9.	10.	11.	12.
$126\,ya^2$	$84\,in^2$	$47.5\,ya^2$	$62\,ya$	$40\,ft^2$	$24\,cm^2$

Chapter 11

Polygons and the Circle

Section 11.2.

Page 321.

Problems

Find the perimeters of the following figures. Be sure to include correct units in your answer:

1. A square.

    28 yd

2. A rectangle.

    11 in, 26 in

3. A triangle.

    18 m, 13 m, 16 m

4. Find the sum of the interior angle measures of a convex 13-gon.

5. Find the sum of the interior angle measures of a convex decagon.

6. The sum of the interior angle measures of a convex polygon is 1080 degrees. How many sides does it have?

7. A wire is bent into the shape of a rectangle with width 5 in and length 7 in. Then, the wire is unbent and reshaped into a triangle. If each side of the triangle has equal length, what is this length?

8.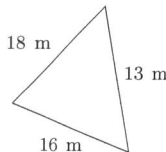

    The perimeter of the rectangle on the left is 136 units. Find the length of $\overline{XY}$.

9. 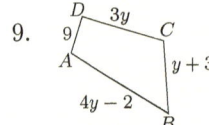 The perimeter of the figure on the left is 74 units. Find the length of $\overline{CD}$.

10. Construct a concave polygon, with a segment having two points inside the polygon and intersecting more than two of its sides.

SOLUTIONS

1. $112\,ya$ 2. $74\,in$ 3. $47\,m$
4. $1980°$ 5. $1440°$ 6. 8 sides
7. $8\,in$ 8. 37 units 9. 24 units 10.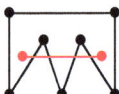

Section 11.3.

Page 324-325.

Problems

1. 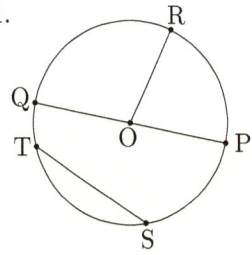 A circle with center O is shown in the figure on the left. Name a diameter, radius, and chord of this circle.

2. 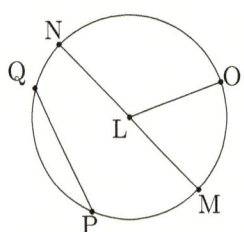 A circle with center L is shown in the figure on the left. If the length of $\overline{LO}$ is 5 units, what is the length of $\overline{MN}$?

In Problems 3 to 8, be sure to include the correct units in your answers. You may use $3.14 \approx \pi$ as your approximation of π.

3. A perfect pizza has a diameter of 16 in. What is its circumference?

4. A bike wheel has a radius of 14 in. What is its circumference?

5. Find the area and circumference of a circle with diameter 9 m.

6. The circumference of a circular painting is 62.8 feet. What is the radius of the painting?

7. The circumference of a circular field is 276.32 yards. What is the area of the field?

8. A rectangular piece of paper with length 36 cm and width 14 cm has two semicircles cut out of it, as shown below. What is the area of the paper that remains?

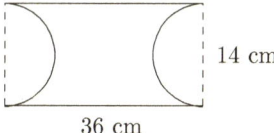

9. A training field is formed by joining a rectangle and two semicircles, as shown below. The rectangle is 90 m long and 72 m wide. Find the area of the training field.

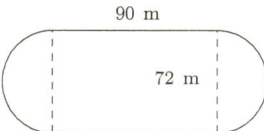

10. An isosceles triangle is placed in a semicircle with a radius of 3 yards, as shown below. Find the area of the shaded region.

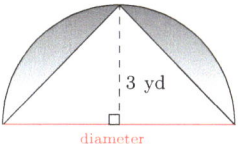

11. At the park there is a pool shaped like a circle. A ring-shaped path goes around the pool. The inner radius of the pool is 11 yards and its outer radius is 17 yards, as shown below. We are going to give a new layer of coating to the path. If one gallon of coating can cover 6 yd^2, how many gallons of coating do we need?

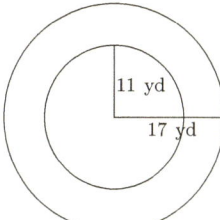

CHAPTER 11. POLYGONS AND THE CIRCLE

SOLUTIONS

1. Diameter: $\overline{QP}$; Radius: $\overline{OR}$; Chord: $\overline{TS}$.
2. 10 units.
3. $50.24\,in.$
4. $87.92\,in.$
5. Area: $63.59\,m^2$ Perimeter: $28.26\,m.$
6. $10\,ft.$
7. $6079.04\,ya^2.$
8. $350.14\,cm^2.$
9. $10549.44\,m^2.$
10. $5.13\,ya^2.$
11. $87.92\,gl.$

Section 11.4.

Page 332-333.

Problems

State whether the two pairs of figures are related through translation, reflection, rotation, or neither:

1.
2.
3.

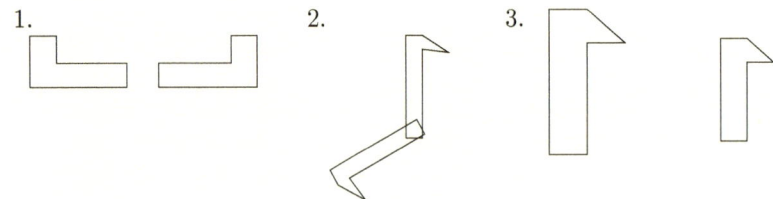

In Problems 4 and 5, draw the figures after the given transformations:

4. Translation 3 units to the left and 1 unit up.

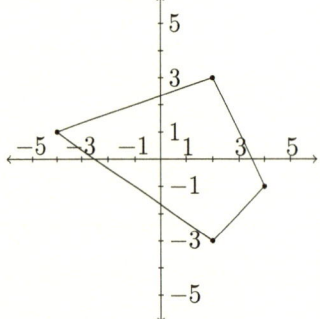

Continuation

5. Reflection across line n.

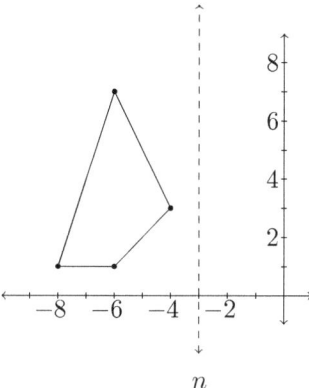

6. Rotation 90 degrees clockwise about the origin.

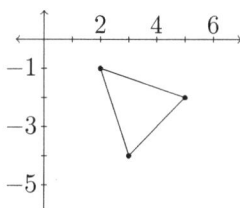

Solve the problems:

7. In the coordinate plane, the point $X = (1, -2)$ is translated to the point $X' = (-2, -3)$. Under the same translation, the points $Y = (-3, 2)$ and $Z = (3, 1)$ are translated to Y' and Z', respectively. What are the coordinates of Y' and Z'?

8. In the coordinate plane, the point $A = (-3, -3)$ is translated to the point $A' = (-4, -2)$. Under the same translation, the points $B = (-7, 0)$ and $C = (0, -6)$ are translated to B' and C', respectively. What are the coordinates of B' and C'?

9. In the coordinate plane, the points $A = (-8, 9)$, $B = (-3, 4)$, and $C = (-1, 5)$ are reflected across the y-axis to the points A', B', and C', respectively. What are the coordinates of A', B', and C'?

10. In the coordinate plane, the points $A = (-11, 10)$, $B = (9, 5)$, and $C = (-4, -3)$ are reflected across the x-axis to the points A', B', and C', respectively. What are the coordinates of A', B', and C'?

SOLUTIONS

1. reflexion.   2. reflexion.   3. neither.

4.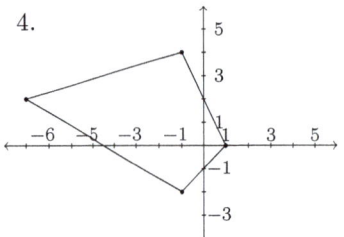

5.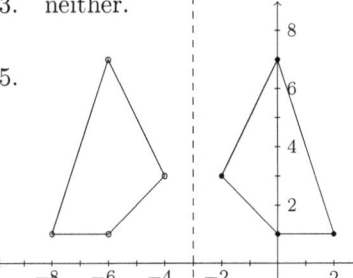

6.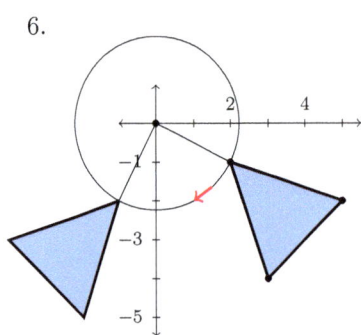

7. $Y' = (-6, 1)$, $Z' = (0, 0)$.

8. $B' = (-8, 1)$, $C' = (-1, -5)$.

9. $A' = (8, 9)$, $B' = (3, 4)$, $C' = (1, 5)$.

10. $A' = (-11, -10)$, $B' = (9, -5)$,

    $C' = (-4, 3)$.

Chapter 12

Three-Dimensional Figures

Section 12.5.

Page 353-354.

> **Problems**
>
> The bases of the prism pictured on the right are regular hexagons. The sides of each hexagon measures 1.5cm, and the altitude of the prism measures 5cm.
>
>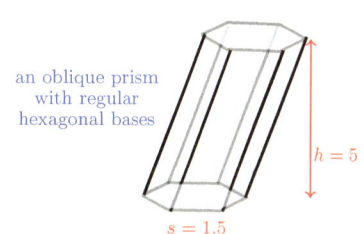
>
> 1. Compute the lateral area of this prism, and also the total area of all the faces.
>
> 2. Compute the volume of the prism.
>
> 3. Instead of a prism, suppose that we have an oblique pyramid. Its height is 7cm, and its base is also a regular hexagon, but with a side measuring 3cm. What is the volume of that oblique pyramid?
>
> 4. What is the surface area of a right cylinder (that is, the area of the tube plus the area of the two faces), with a height equal to 6cm and with the radius of the two faces equal to 5cm?

Continuation

The right cylinder pictured by the figure can be filled with water. The stopper, or cork, is a sphere which goes half way into the cylinder, so that the lower hemisphere of the stopper would be submerged into the water.

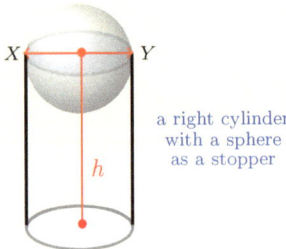

a right cylinder with a sphere as a stopper

The cylinder together with its stopper as in the figure, form a solid, which is made of the base of the cylinder, the wall of the cylinder, and the upper hemisphere of the sphere. The segment $\overline{XY}$ is both a diameter of the cylinder, and a diameter of the sphere. Its length is equal to 6cm. The height h of the cylinder is equal to 10cm.

5. What is the surface area of the sphere, and what is its volume?

6. Compute the lateral area of the cylinder, and also the total area of the solid formed by the cylinder and the stopper.

7. Compute the volume of the water filling the cylinder with its stopper. (This is the part of the volume of the cylinder, which does not include the stopper.)

8. Compute the lateral area of a right cone with slant height equal to 8cm and radius equal to 3 cm. Compute also the surface area of this cone.

9. Compute the volume of a right cone with a radius equal to 7cm, and a height equal to 11cm.

10. Compute the volume of a sphere with a radius equal to 15cm, as the difference between:

    (a) a cylinder having its diameter $2r = 30$ equal to its height,

    (b) and the combined volume of two cones, each with a height equal to r and a radius equal to $2r$.

(If needed, go back to Figure 12.18 on page 352 of Volume I.)

11. Unfolding of the wall of a right cylinder produces a rectangle. (See Figure 12.9 on page 341 of Volume I.) Suppose that one unfolds the wall of an oblique cylinder in a similar manner. Which geometrical figure would result of such an unfolding? What would be its surface area if the height of the cylinder is 8 cm and its radius 3cm?

SOLUTIONS

1. $\begin{cases} 45\,cm^2; \\ 56.69\,cm^2. \end{cases}$ 2. $29.23\,cm^3$.   3. $54.56\,cm^3$.

4. $345.4\,cm^2$. 5. $\begin{cases} \text{Surface area: } 113.04\,cm^2; \\ \text{Volume: } 113.04\,cm^3. \end{cases}$ 6. $\begin{cases} \text{Lateral area: } 188.4\,cm^2; \\ \text{Total area}: 244.92\,cm^2. \end{cases}$

7. $226.08\,cm^3$ 8. $\begin{cases} \text{Lateral area: } 75.36\,cm^2 \\ \text{Total area}: 103.62\,cm^2 \end{cases}$ 9. $564.15\,cm^3$.

10. The computation is based on the picture below reproducing Figure 12.18.

$$\pi r^2 \times 2r \quad - \quad 2 \times \left(\frac{1}{3}\pi r^2 \times r\right) \quad = \frac{4}{3}\pi r^3$$

$$21,195\,cm^3 \quad - \quad 7,065\,cm^3 \quad = 14,130\,cm^3$$

Chapter 13

Probability

Section 13.2.

Page 358.

> **Problem**
>
> Linda must choose a size, a flavor, and a topping for her ice cream dessert. There are two sizes to choose from, two flavors to choose from, and two toppings to choose from. The tree diagram below shows the possible outcomes. Use the diagram to answer the following questions:
>
> 1. How many outcomes are there?
> 2. How many outcomes have both strawberry and nuts chosen?
> 3. How many outcomes do not have chocolate being chosen?
>
>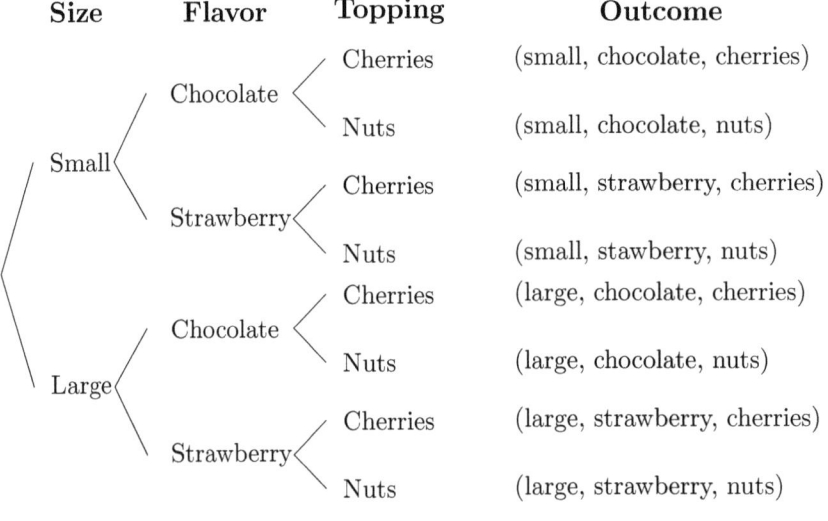

123

SOLUTIONS

1. 8 outcomes.   2. 2 oucomes.   3. 4 outcomes.

Section 13.4.

Page 367-368.

> **Problems**
>
> Solve the word problems. Write your answers as fractions in simplest form, if applicable:
>
> 1. Suppose that the genders of the three children of a family are soon to be revealed. An outcome is represented by a string such as GBB, in this case meaning that the oldest child is a girl, the second oldest is a boy, and the youngest is a boy. What is the sample space? You should write it in the form $\{GGG, \ldots\}$.
>
> 2. (Continuation.) For each of the three events in the table, write out the outcome(s) that are contained in the event. In the last column, enter the probability of the event. You may assume that, for each of the children, the probability that the child is a girl is $\frac{1}{2}$.
>
> | Event | Outcomes | Probability |
> |---|---|---|
> | Two or more girls | | |
> | Exactly one girl | | |
> | A girl on each of the last two births | GGG, BGG | $\frac{1}{4}$ |
>
> 3. A spinner with 10 equally sized slices is shown below. The dial is spun and stops on a slice at random. What is the probability that the dial stops on a yellow slice?
>
>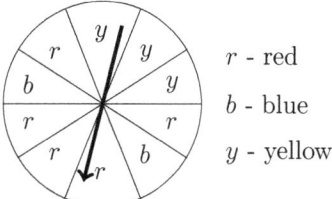
>
> r - red
> b - blue
> y - yellow

CHAPTER 13. PROBABILITY

Continuation. Solve the word problems 4-8 below.

Write your answers as fractions in simplest form, if applicable.

4. A spinner with 8 equally sized slices is shown below. The dial is spun and stops on a slice at random. What is the probability that the dial stops on a red slice?

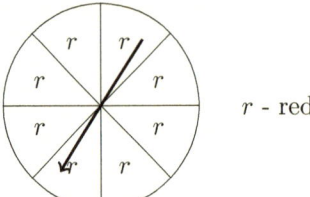

r - red

5. A box is filled with 6 brown cards, 6 green cards, and 2 blue cards. A card is chosen at random from the box. What is the probability that the card is *not* green? In this problem and in the next two problems, you may assume that all the cards have the same probability of being chosen.

6. A box is filled with 7 green cards, 5 brown cards, and 3 blue cards. A card is chosen at random from the box. What is the probability that the card is a green or blue card?

7. A bag contains 10 cards numbered 1 through 10. A card is randomly chosen from the bag. What is the probability that the card has a multiple of 3 on it?

8. A class is made up of 10 boys and 8 girls. Half of the boys wear glasses. A student is selected at random form the class. What is the probability that the student is a boy with glasses? All the students are selected with the same probability.

SOLUTIONS

1. $\{GGG, GGB, GBG, GBB, BGG, BGB, BBG, BBB\}$.

2.
Event	Outcomes	Probability
Two or more girls	GGG, BGG, GGB	$\frac{1}{2}$
Exactly one girl	GBB, BGB, BBG	$\frac{3}{8}$
A girl on each of the last two births	GGG, BGG	$\frac{1}{4}$

3. $\frac{3}{10}$.   4. 1.   5. $\frac{4}{7}$.

6. $\frac{2}{3}$.   7. $\frac{3}{10}$.   8. $\frac{5}{18}$.

Section 13.6.

Page 373-374.

> **Problems**
>
> 1. Let A and B be two independent events. We suppose that their probabilities are neither equal to 0 nor equal to 1. Give a formula for the probability of each of the following events:
>
>   (a) neither the event A nor the event B occurs;
>
>   (b) A occurs but B does not occur;
>
>   (c) either A does not occur or B does not occur;
>
>   (d) either A occurs or B does not occur.
>
> 2. If A and B are independent events, what can be said about their complements $\overline{A}$ and $\overline{B}$?
>
> 3. For each of the following, check whether the two events are independent and prove your argument:
>
>   (a) in three tosses of a fair coin, the events 'at least two tosses are heads' and 'at least one toss is a tail';
>
>   (b) in a roll of two fair dice, the events 'the sum of the dice is odd' and 'the second die is a multiple of 3'.
>
> **Hint.** The percentages given in some of the problems should be regarded as probabilities. In Problem 4, for example, the sentence "**A certain HIV test detects the virus in 98% of infected subjects**" must be interpreted as "**The conditional probability that a certain HIV test detects the virus, given that the subject is infected, is .98**".
>
> 4. A certain HIV test detects the virus in 98% of infected subjects. The test also classifies as infected about 0.8% of all noninfected subjects. If 0.5% of the population is infected with HIV, what is the probability that a person who tests positive actually has the virus?

Continuation

5. At a certain college, 45% of the students are female, and 19% of the students major in civil engineering. Furthermore, 8% of the students are both female and major in civil engineering.

    (a) What is the probability that a randomly selected female student majors in civil engineering?

    (b) What is the probability that a randomly selected civil engineering major is female?

6. A survey showed that 23% of college students read newspapers on a regular basis and that 83% of them regularly watch the news on TV. The survey also showed that 21% of college students both follow TV news regularly and read newspapers regularly.

    (a) What is the probability that a randomly selected college student reads newspapers regularly, given that he or she watches TV news regularly?

    (b) What is the probability that a student watches TV news regularly, given that he or she regularly reads newspapers?

7. Susan and Archie decide to play the following game of dice. Susan always plays first and tosses the die. Suppose that the result is i, with i in $\{1,\ldots,6\}$. Then Archie picks the die and plays. If the result is smaller than i, he is supposed to continue playing until he gets a result $j \geq i$. If $j = i$, Archie loses \$2, which he gives to Susan. Otherwise, he wins and receives \$$j - i$. You may assume that the die is fair. Suppose that they play that game just once.

    (a) Construct an appropriate sample space for that situation.

    (b) Construct the probability distribution.

    (c) Compute the probability that Archie: (1) wins; (2) wins \$2.

    (d) Are the events 'Susan's result is even' and 'Archie's result is odd' independent?

    (e) Which player would you rather be? Justify your choice.

SOLUTIONS

1. (a) $\mathbb{P}(\bar{A} \cap \bar{B})$; (b) $\mathbb{P}(A \cap \bar{B})$; (c) $\mathbb{P}(\bar{A} \cup \bar{B})$; (d) $\mathbb{P}(A \cup \bar{B})$.

2. The events $\bar{A}$ and $\bar{B}$ are also independent, because:

$$\mathbb{P}(\bar{A})\mathbb{P}(\bar{B}) = (1 - \mathbb{P}(A))(1 - \mathbb{P}(B)) = 1 - \mathbb{P}(A) - \mathbb{P}(B) + \mathbb{P}(A \cap B)$$
$$= \mathbb{P}(\bar{A} \cap \bar{B}).$$

Indeed, writing Ω for the sample space, we have

$$1 = \mathbb{P}(\Omega) = \mathbb{P}(\bar{A} \cap \bar{B}) + \mathbb{P}(A) + \mathbb{P}(B) - \mathbb{P}(A \cap B).$$

(Examine Figure 13.6 on page 367. The green part pictures $\mathbb{P}(\overline{A \cup B}) = \mathbb{P}(\bar{A} \cap \bar{B})$.)

3. (a) These two events are not independent. With h for 'head' and t for tail, we write the sample space as: $\{hhh, hht, hth, thh, tth, tht, htt, ttt\}$.

The two events are

at least two tosses are heads: $H_2 = \{hhh, hht, hth, thh\}$
at least one toss is a tail: $T_1 = \{hht, hth, thh, tth, tht, htt, ttt\}$

This yields: $\mathbb{P}(H_2 \cap T_1) = \mathbb{P}\{hht, hth, thh\} = \dfrac{3}{8}$,

$$\mathbb{P}(H_2) \times \mathbb{P}(T_1) = \dfrac{4}{8} \times \dfrac{7}{8} = \dfrac{28}{64} = \dfrac{3.5}{8} \neq \dfrac{3}{8}.$$

(b) With the sample space: $\{(i,j) \mid 1 \leq i \leq 6, 1 \leq j \leq 6\}$,
the two events are:

$S_{odd} = \{(1,2),(1,4),(1,6),(2,1),(2,3),(2,5),(3,2),(3,4),(3,6),$
$\qquad (4,1),(4,3),(4,5),(5,2),(5,4),(5,6),(6,1),(6,3),(6,5)\},$
$M_3 = \{(1,3),(1,6),(2,3),(2,6),(3,3),(3,6),(4,3),$
$\qquad (4,6),(5,3),(5,6),(6,3),(6,6)\}.$

So, we have $|S_{odd}| = 18$ and $|M_3| = 12$, and

$S_{odd} \cap M_3 = \{(1,6),(2,3),(3,6),(4,3),(5,6),(6,3)\}, \qquad |S_{odd} \cap M_3| = 6.$

This yields

$$\mathbb{P}(S_{odd}) \times \mathbb{P}(M_3) = \dfrac{18}{36} \times \dfrac{12}{36} = \dfrac{1}{2} \times \dfrac{1}{3} = \dfrac{1}{6},$$
while $\qquad \mathbb{P}(S_{odd} \cap M_3) = \dfrac{6}{36} = \dfrac{1}{6}.$

The two events are thus independent.

4. We define the events
$$\begin{cases} I & \text{the individual is infected,} \\ \bar{I} & \text{the individual is not infected;} \end{cases} \quad \begin{cases} D & \text{HIV is detected,} \\ \bar{D} & \text{HIV is not detected.} \end{cases}$$

The data of the problem mean that we have:
$$\mathbb{P}(D\,|\,I) = .98, \quad \mathbb{P}(D\,|\,\bar{I}) = .08, \quad \mathbb{P}(I) = .05.$$

The problem asks: what is the value of $\mathbb{P}(I\,|\,D)$? We have successively:
$$P(\bar{I}) = 1 - \mathbb{P}(I) = .95,$$

and so, by Theorem 13.A,
$$\mathbb{P}(D) = \mathbb{P}(D\,|\,I)\mathbb{P}(I) + \mathbb{P}(D\,|\,\bar{I})\mathbb{P}(\bar{I})$$
$$= .98 \times .05 + .08 \times .95 = .125.$$

We now have
$$\mathbb{P}(I \cap D) = \mathbb{P}(D\,|\,I)\mathbb{P}(I)$$
$$= .98 \times .05 = .049,$$

which yields
$$\mathbb{P}(I\,|\,D) = \frac{\mathbb{P}(I \cap D)}{\mathbb{P}(D)} = \frac{.049}{.125} \approx .392.$$

5. In the style of the previous problem, we define
$$\begin{cases} F & \text{the individual is a female,} \\ \bar{F} & \text{the individual is a male;} \end{cases}$$
$$\begin{cases} CE & \text{the individual majors in civil engeneering,} \\ \overline{CE} & \text{the individual does not major in civil engeneering.} \end{cases}$$

Using these notation, the problem state the following:
$$\mathbb{P}(F) = .45, \quad \mathbb{P}(CE) = .19, \quad \mathbb{P}(F \cap CE) = .08.$$

(a) $\mathbb{P}(CE\,|\,F) = \frac{\mathbb{P}(F \cap CE)}{\mathbb{P}(F)} = \frac{.08}{.45} \approx .178.$

(b) $\mathbb{P}(F\,|\,CE) = \frac{\mathbb{P}(F \cap CE)}{\mathbb{P}(CE)} = \frac{.08}{.19} \approx .421.$

6. We omit the solution of this problem.

7. Let (i, j) be any sample point, with i and j denoting Susan and Archie's results, respectively.

   (a) The sample space is $\Omega = \{(i, j)\,|\,1 \leq i \leq 6, i \leq j \leq 6\}$.

   (b) The outcome probabilities are displayed in the Table A below. The resulting gains of Archie are displayed in Table B.

Susan's result

	1	2	3	4	5	6
1	$\frac{1}{6} \times \frac{1}{6}$	$\frac{1}{6} \times \frac{1}{6}$	$\frac{1}{6} \times \frac{1}{6}$	$\frac{1}{6} \times \frac{1}{6}$	$\frac{1}{6} \times \frac{1}{6}$	$\frac{1}{6} \times \frac{1}{6}$
2	X	$\frac{1}{6} \times \frac{1}{5}$	$\frac{1}{6} \times \frac{1}{5}$	$\frac{1}{6} \times \frac{1}{5}$	$\frac{1}{6} \times \frac{1}{5}$	$\frac{1}{6} \times \frac{1}{5}$
3	X	X	$\frac{1}{6} \times \frac{1}{4}$	$\frac{1}{6} \times \frac{1}{4}$	$\frac{1}{6} \times \frac{1}{4}$	$\frac{1}{6} \times \frac{1}{4}$
4	X	X	X	$\frac{1}{6} \times \frac{1}{3}$	$\frac{1}{6} \times \frac{1}{3}$	$\frac{1}{6} \times \frac{1}{3}$
5	X	X	X	X	$\frac{1}{6} \times \frac{1}{2}$	$\frac{1}{6} \times \frac{1}{2}$
6	X	X	X	X	X	$\frac{1}{6} \times 1$

(Archie's result: rows 1–6)

Table A: Outcome probabilities.

Susan's result

	1	2	3	4	5	6
1	-2	1	2	3	4	5
2	X	-2	1	2	3	4
3	X	X	-2	1	2	3
4	X	X	X	-2	1	2
5	X	X	X	X	-2	1
6	X	X	X	X	X	-2

(Archie's result: rows 1–6)

Table B: Archie's gains in dollars.

(c) and (e). The probability that Susan wins is the probability that Archie's result is equal to Susan result, namely, the sum of the six probabilities printed in red in the diagonal cells of Table A:

$$\mathbb{P}\{(i,j) \mid i = j, 1 \leq i \leq 6\} = \frac{1}{36} + \frac{1}{30} + \frac{1}{24} + \frac{1}{18} + \frac{1}{12} + \frac{1}{6} \approx .408\,.$$

So, the probability that Archie's wins is approximately equal to $1 - .408 = .592$. That answers also question (e): it is preferable to play at Archie's place in this game.

To find the probability that Archie wins exactly 2 dollars, we have to sum the probabiities of all the cells marked '2' in Table B. These probabilities are printed in blue in Table A. This probability is:

$$\mathbb{P}\{(i,j) \mid i = j+2, 2 \leq i \leq 6\} = \tfrac{1}{6} \times \tfrac{1}{6} + \tfrac{1}{5} \times \tfrac{1}{6} + \tfrac{1}{4} \times \tfrac{1}{6} + \tfrac{1}{3} \times \tfrac{1}{6} = .158\overline{3}.$$

(d) Writing S_E for 'Susan's result is even' and A_O for 'Archie's result is odd', these events are independent if

$$\mathbb{P}(S_E \cap A_O) = \mathbb{P}(S_E) \times \mathbb{P}(A_O).$$

Computing these three probabilities from Table A, we get:

$$\mathbb{P}(S_E) = \mathbb{P}\{(i,j) \mid i = 2, 4, 6;\ 1 \leq j \leq 6\}$$
$$= \frac{1}{6} \times \left(\frac{3}{6} + \frac{3}{5} + \frac{2}{4} + \frac{2}{3} + \frac{1}{2} + 1 \right) \approx .63\,.$$

$$\mathbb{P}(A_O) = \frac{1}{6} + \frac{1}{6} + \frac{1}{6} = \frac{1}{2}.$$

$$\mathbb{P}(S_E) \times \mathbb{P}(A_O) \approx .63 \times .5 = .315$$

$$\mathbb{P}(S_E \cap A_O) = \mathbb{P}\{(i,j) \mid i = 2, 4, 6; j = 1, 3, 5\}$$
$$= \frac{3}{36} + \frac{2}{24} + \frac{1}{12} = \frac{3}{12} = .25$$

So, the fact that $.25 \neq .315$ means that the events S_E and A_O are not independent.

Page 378.

> **Problems**
>
> 1. Arnold has four coins in his pocket: two standard coins, a coin with heads on both sides, and a coin with tails on both sides. He pulls one coin out of his pocket, looks at one side of the coin, and notices that it is a head. What is the probability that the other side is a tail?

2. Five prisoners A, B, C, D, and E are informed that three of them are going to be released. Prisoner B is told that either C or D is going to be released. What is the probability that prisoner B is going to be released, assuming that the guard is telling the truth?

3. There are three hats on the table: two green and one red. There are also two belts in the closet: one is green and the other one is red. Each of two sisters, Audrey and Susan, is randomly picking one hat and one belt.

   (a) What is the probability that each of the two sisters has matching colors?

   (b) What is the conditional probability that Audrey has a green hat, given that Susan has matching colors?

4. Arnold must travel to a foreign country where an epidemic of a deadly strain of Asian flu has struck. About 15% of the people in large urban centers have died already. The chances of dying, for a nonvaccinated person in this country, have been evaluated at 1/6. Due to the fast development of the disease, only 25% of the people have been vaccinated. Arnold wants to know (as accurately as possible) what his chances of survival are if he undergoes vaccination before departure. Can you help him?

5. Suppose that, in some country, the probability of having a girl is .5. Take one family at random and suppose that they have two children, and that one of their children is a girl. What is the probability that the other child is a boy?

6. Archie is into traveling again. This time, he is stranded in a war zone in the faraway land of the Xhosas. His only chance of survival is to make it to Sudland, which is 100 miles away, and can be reached either by crossing a desert or by going through the mountains. Both are very chancy enterprises. Only 38% of the people who try to make it to Sudland succeed. Archie knows that about $\frac{4}{5}$ who attempt to reach Sudland try to cross the desert. It is also known that the probability of reaching Sudland by the mountain route is 90%. Archie wants to know what his chances of survival are if he goes through the desert. Can you help him?

SOLUTIONS

1. Intuitively, the probability that the other side of the coin is a tail should be $\frac{2}{3}$. Indeed, since the face of the coin that Arnold looked at is a head, he knows that it can't be the 2-tail coin. Two of the three remaining coins are standard. The other face of Arnold's coin is a tail only if it is one of the those two. We can check whether our intuition is confirmed by a formal analysis of the type used in the solution to Problem 2 below.

2. We write ABC to mean: 'Prisoners A, B, and C are going to be released', with a similar notation for the other cases. Our sample space is

$$\{ABC, ABD, ABE, ACD, ACE, ADE, BCD, BCE, BDE, CDE\},$$

with each of the 10 outcomes having the same $\frac{1}{10}$ probability. The event that Prisoner B is going to be released is

$$B = \{ABC, ABD, ABE, BCD, BCE, BDE\} \qquad (\star)$$

which happens with probability

$$\mathbb{P}(B) = \mathbb{P}(ABC, ABD, ABE, BCD, BCE, BDE) = \frac{6}{10}.$$

We write '$C \cup D$' for the event: 'either C or D is going the be released'; we have thus

$$C \cup D = \overbrace{\{ABC, ACD, ACE, BCD, BCE, CDE\}}^{C}$$
$$\cup \underbrace{\{ABD, ACD, ADE, BCD, BDE, CDE\}}_{D}. \qquad (\star\star)$$

This event occurs with all outcomes except one: the outcome ABE. So, we have

$$\mathbb{P}(C \cup D) = 1 - \mathbb{P}(ABE) = \tfrac{9}{10}.$$

The conditional probability that B is going to be released, given that C or D is going to be released is thus

$$\mathbb{P}(B \mid C \cup D) = \frac{\mathbb{P}(B \cap (C \cup B))}{\mathbb{P}(C \cup D)} = \frac{\mathbb{P}((B \cap C) \cup (B \cap D))}{\mathbb{P}(C \cup D)}$$

and by $(\star)$ and $(\star\star)$,

$$= \frac{\mathbb{P}(\{ABC, BCD, BCE\} \cup \{ABD, BCD, BDE\})}{\mathbb{P}(C \cup D)}$$

$$= \frac{\mathbb{P}\{ABC, BCD, BCE, ABD, BDE\}}{\mathbb{P}(C \cup D)} = \frac{\frac{5}{10}}{\frac{9}{10}} = \frac{5}{9}.$$

3. We omit the solution of this problem and that of Problem 5.

4. We write

V	for	'vaccinated'
$\bar{V}$	for	'not vaccinated'
D	for	'a death due to the disease'.

From Theorem 13.A, we have:

$$\mathbb{P}(D) = \mathbb{P}(D \mid V)\mathbb{P}(V) + \mathbb{P}(D \mid \bar{V})\mathbb{P}(\bar{V})$$

which using the data of the problem translates into

$$.15 = \mathbb{P}(D\,|\,V) \times .25 + \frac{1}{6} \times .75\,.$$

Solving for $\mathbb{P}(D\,|\,V)$, we obtain

$$\mathbb{P}(D\,|\,V) = \frac{.15 - \frac{.75}{6}}{.25} = .1\,.$$

(This is still a high risk for Arnold.)

6. We write: D for   'crossing the desert'
                      M for   'going through the mountains'
                      S for   'reaching Sudland'.

With these notation and applying Theorem 13.A, we get

$$\mathbb{P}(S) = \mathbb{P}(S\,|\,D)\mathbb{P}(D) + \mathbb{P}(S\,|\,M)\mathbb{P}(M)\,.$$

Using the data of the problem, this equation translates into:

$$.38 = \mathbb{P}(S\,|\,D) \times \frac{4}{5} + .9 \times \frac{1}{5}\,.$$

Solving for $\mathbb{P}(S\,|\,D)$, we obtain

$$\mathbb{P}(S\,|\,D) = \left(.38 - .9 \times \frac{1}{5}\right) \times \frac{5}{4} = .25\,.$$

So, Archie has a 25% chance of reaching Sudland by crossing the desert. By all means, he should take the mountain route.

Chapter 14

Statistics and Data Analysis

Section 14.1.

Page 385.

> **Problems**
>
> Draw the histograms and the cumulative frequency distributions for the three data below:
>
> 1. The shopping times (in minutes) of a sample of 17 shoppers at a local clothing store are:
>   36, 45, 25, 47, 23, 44, 28, 48, 22, 40, 33, 37, 38, 27, 26, 39, 24.
>   Label your y-axis as frequency and scale it from 0 to .25, .5, .75, 1.
>   Label your x-axis as time and scale it into 4 intervals of
> $22-28, 29-35, 36-42$, and $43-49$.
>
> 2. Here are the scores for 17 students on a history test:
>   78, 69, 92, 87, 86, 76, 72, 70, 84, 77, 85, 89, 71, 83, 73, 75, 90.
>   Label your y-axis as frequency and scale it from 0 to .25, .5, .75, 1.
>   Label your x-axis as results and scale it into 4 intervals of
> $69-74, 75-80, 81-86$, and $87-92$.
>
> 3. The heights (in inches) of a sample of 19 adult males are:
>   64, 76, 77, 63, 78, 68, 66, 80, 72, 65, 80, 73, 69, 70, 71, 67, 75, 79, 74.
>   Label your y-axis as frequency and scale it from 0 to .25, .5, .75, 1.
>   Label your x-axis as height and scale it into 4 intervals of
> $61-65, 66-70, 71-75$, and $76-80$.

Solutions

1. Histogram (the green bars) and cumulative frequency distribution (the yellow bars) of the shopping times (in minutes) of 17 shoppers at a local clothing store.

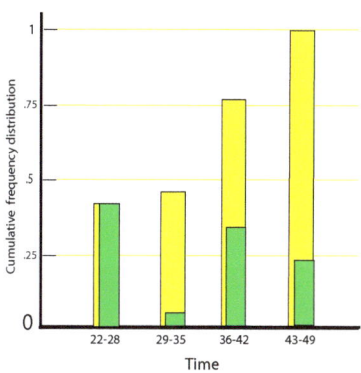

2. Histogram and cumulative frequency distribution of the scores of 17 students on a history test.

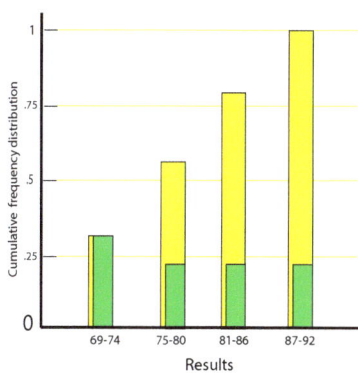

3. Histogram and cumulative frequency distribution of the heights (in inches) of 19 adults male.

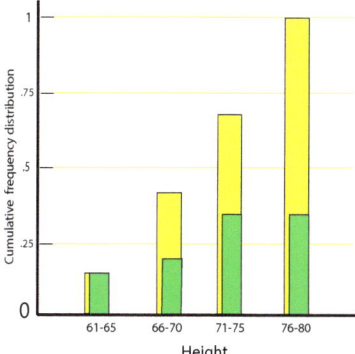

Section 14.2.

Page 392.

> **Problems**
>
> Compute the mean and the median for the three sets of data below[a].
>
> 1. The shopping times (in minutes) of a sample of 17 shoppers at a local clothing store:
>   36, 45, 25, 47, 23, 44, 28, 48, 22, 40, 33, 37, 38, 27, 26, 39, 21.
>
> 2. The scores for 17 students on a history test:
>   78, 69, 92, 87, 86, 76, 72, 70, 84, 77, 85, 89, 71, 83, 73, 75, 90.
>
> 3. The heights (in inches) of a sample of 19 adult males:
>   64, 76, 77, 63, 78, 68, 66, 80, 72, 65, 80, 73, 69, 70, 71, 67, 75, 79, 74.
>
> ---
> [a]They are the same data as those of the preceding set of problems.

SOLUTIONS

1. The mean shopping times of the 17 shoppers is $\frac{579}{17} \approx 34.059$. The number of observations is odd. So, we are in Case 1 of the computation of the median of a distribution. Ordering the observations, we get

$$\underbrace{21 < 22 < \ldots < 33}_{\text{8 observations}} < 36 < \underbrace{37 < 38 < \ldots < 48}_{\text{8 observations}}.$$

So the median is 36.

2. We omit the answer to this problem.

3. The mean height of the 19 adult males is: $\frac{1367}{19} \approx 71.947$. The number of observations is odd, as in Problem 1. Ordering the observations yields:

$$\underbrace{63 < 64 < \ldots < 71}_{\text{9 observations}} < 72 < \underbrace{73 < 74 < \ldots < 80}_{\text{9 observations}}.$$

So the median is 72.

Page 396-397.

Problems

Find the mode for each of the following data sets:

1. The number of siblings for each of 12 students:
   $4, 1, 0, 0, 2, 2, 2, 2, 1, 3, 4, 4$.

2. The number of times each of 15 people ate out last month:
   $4, 3, 4, 7, 5, 4, 6, 4, 4, 6, 6, 6, 5, 5, 5$.

3. Twenty people are currently married. The number of previous divorces are: $0, 0, 1, 2, 0, 4, 5, 0, 0, 1, 3, 2, 4, 2, 0, 1, 1, 0, 2, 3$.

Find the mean and median of the following data sets:

4. A group of 5 students was asked, "How many hours did you watch television last week?". Here are their responses: $14, 20, 11, 18, 16$.

5. Each of the 7 cats in a pet store was weighed. Here are their weights (in pounds): $8, 9, 15, 8, 11, 10, 7$.

Solve the word problems:

6. Kala is participating in a 6-day cross-country biking challenge. She biked for 52, 58, 57, 50, and 61 miles on the first five days. How many miles does she need to bike on the last day so that her average is 58 miles per day?

7. Scott has scored 62, 70, and 77 on his previous three tests. What score does he need on his next test so that his average is 76?

8. Mary has scored 34, 26, 29, 21, and 29 points in her five basketball games so far. How many points does she need to score in her next game so that her average is 28 points per game?

Describe how both the mean and median will change if the following changes are made to each of the data sets in Problems 9, 10 and 11:

9. The numbers of students in the schools in a district are given here: $216, 259, 271, 304, 323, 374, 381$. Suppose that the number 381 from this list changes to 500.

> **Continuation**
>
> 10. The weekly salaries (in dollars) for 9 employees of a small business are given here: $577, 610, 630, 659, 670, 737, 799, 843, 910$. Suppose that the \$670 salary changes to \$725.
>
> 11. The monthly rents (in dollars) paid by 8 people are given here: $860, 945, 960, 980, 990, 1000, 1055, 1090$. Suppose that one of the people moves, and his rent changes from \$860 to \$940.
>
> Answer the questions below:
>
> 12. The following numbers of people attended the last 9 screenings of a movie: $195, 199, 202, 203, 204, 205, 207, 208, 211$. Which measure should be used to summarize the data?
>
> 13. In a survey, a soft drink company asks people to name as many brands of soft drink as they can. Which measure gives the most frequently mentioned brand?
>
> 14. The last 9 calls to a customer support line had the following lengths (in minutes): $22, 23, 24, 25, 27, 29, 30, 31, 63$. Which measure should be used to summarize the data?

SOLUTIONS

Some of these problems are quite simple. We omit the responses to Problems 4, 5, 7, 8, 11, and 13.

1. The distribution of the number of siblings is unimodal.
   The mode of the distribution is 2.

2. Five of these people ate out 4 times last month. The mode is 4.

3. The distribution of the number of divorces among these 20 people is unimodal. Seven people did not divorce. So, the mode is 0.

6. Writing x for the number of miles biked by Kala on the last day, the value of x should be such that
$$\frac{52 + 58 + 57 + 50 + 61 + x}{6} = 58$$
miles per day on the average. Solving for x, we get
$$x = 58 \times 6 - (52 + 58 + 57 + 50 + 61)$$
$$= 70.$$

So, Kala must bike exactly 70 miles on her last day to achieve her goal.

9. The mean number of students is:
$$\frac{216 + 259 + 271 + 304 + 323 + 374 + 381}{7} = 304,$$
which is also the median of these numbers. Replacing 381 by 500 will not change the median, but the mean will become 321.

10. This problem is similar to Problem 9, in that the number of outcomes is odd. The median before the change is the 5th oucome in the order
$$577 < 610 < 630 < 659 < 670 < 737 < 799 < 843 < 910,$$
that is, 670$. Before the change, the mean is
$$\frac{577+610+630+659+670+737+799+843+910}{9} = 715.$$
However, since 670 was the median, changing 670 into 725 changes the median, which is now 725, while the mean becomes $721.\bar{1}$.

12. The numbers of people attending the last 9 screening of the movie are very close together. Either the mean or the median would be an appropriate summary of the data.

14. Here however, the numbers are also close together, except for the outlier: the 63 minutes taken by one customer. Using the mean to summarize the data would be misleading. The median is the appropriate measure here.

Section 14.4

Page 409-410.

Problems

1. In each of Case (a) and (b) in the table below, compute the quartiles and the interquartile ranges, and draw the box-and-whiskers plots.

	\multicolumn{10}{c}{Possible Outcomes}
	1	2	3	4	5	6	7	8	9	10
(a) Number of observed outcomes	8	6	10	19	16	30	25	20	17	12
(b) Number of observed outcomes	5	10	36	53	50	64	35	22	15	3

2. **For the very curious students.** For the same Cases (a) and (b) data as in the table of Problem 1, draw the line graphs of the cumulative distributions, and compute the quartiles and the interquartile ranges by that method. Compare the results with those obtained by the method to be used in Problem 1.

3. Draw the scatter plots for each of the sets of ordered pairs (a) and (b) below.

   (a) = {(2, 10), (2, 12), (3, 5), (3, 7), (3, 12), (3, 20), (4, 10), (4, 15), (4, 22), (4, 25), (5, 22), (5, 26), (5, 31), (5, 35), (6, 20), (6, 25), (6, 32), (6, 40), (6, 43), (7, 25), (7, 29), (7, 35), (7, 52), (8, 33), (8, 40), (8, 45), (8, 55), (9, 43), (9, 48), (10, 45), (10, 55), (10, 60)}.

   (b) = {(1, 124), (1, 129), (1, 140), (2, 110), (2, 112), (2, 125), (3, 100), (3, 102), (4, 80), (4, 94), (5, 80), (6, 75), (6, 82), (6, 84), (7, 79), (8, 88), (8, 92), (9, 95), (9, 105), (10, 102), (10, 104), (10, 109), (10, 118)}.

4. Study the scatter plot that you made for the ordered pairs in 3.(a) above. By trial-and-error, try to find the best linear equation that could be a good summary for these data. This linear equation has the form $y = ax + b$. What could be the values of a and b? Would it make sense to summarize the scatter plots of the order pairs in 3.(b) by a linear equation?

5. **For the very curious students.** Study the scatter plot that you made for the ordered pairs in 3.(b) above. Could you find a parabola, that is, a curve of described by the equation

$$y = ax^2 + bx + c$$

that would be a good summary of these data? Think about how we could find such a parabola by some computation. (The answer is part of a more advanced course.)

SOLUTIONS

1. (a) Box-and-whiskers plot.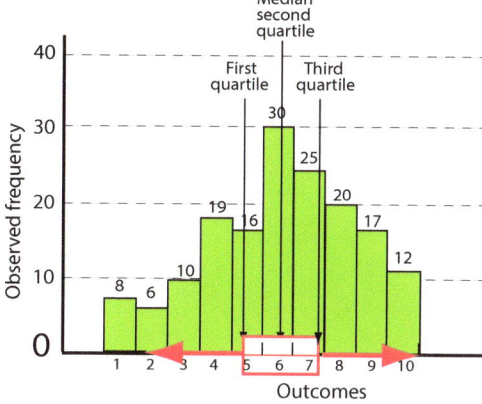

2. (a)

   The line graph of the cumulative distribution for the Case (a) of Problem 1, with the quartiles and the interquartile range.
   The placement of the quartiles is more precise, especially if we use the method described for the curious students on page 403.

   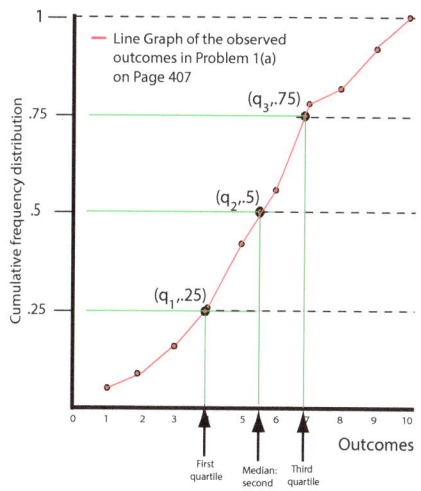

3. (a) and 4.

Scatter plot of the set of ordered pairs in Problem 3(a), with a line representing the linear equation $y = 7x - 9$ obtained by trial-and-error. In a more advanced statistics course, such a line would be obtained by minimizing the distances between the ordinates y_i of the points (x_i, y_i) and the the corresponding values $y_i = ax_i + b$. Here, we have $a = 7$ and $b = -9$.

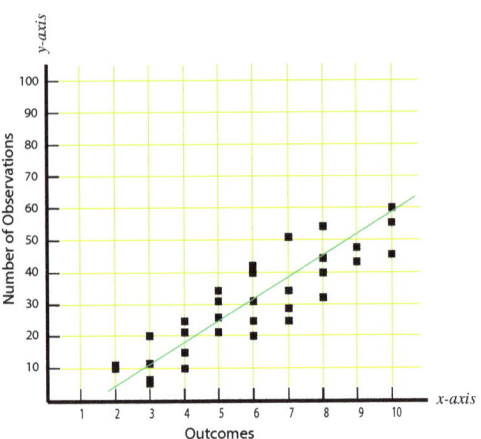

4. and 5.

Scatter plot of the set of ordered pairs in Problem 3(b). It would not make sense to represent such data by a straight line. A parabola would be more appropriate. In standard methods, such a parabola would be obtained by estimating the values of the parameters a, b and c in the equation $y = ax^2 + bx + c$ by a minimization of the squares of the distances between the y_i values of the observations (x_i, y_i) and the corresponding values $ax_i^2 + bx_i + c$. This approach is referred to as the *least squares* method.

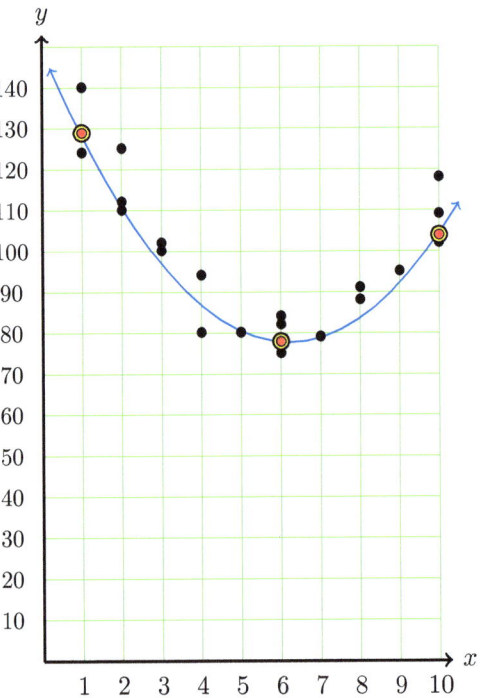

We fitted the parabola pictured above by requiring that it contained the three points marked by yellow circles with a red center in the above graph. This leads to a linear system of three equations with the three unknowns a, b and c. Solving this system gave the equation of the parabola. Needless to say, this is not the best method to fit a parabola to a scatter plot. The standard tool is the *least square* method, which proceed by finding the the values of a, b and c by minimizing the vertical distances of the parabola to the points of the diagram.

Appendix. Abbreviations of the measurement units.

Units of the measurement of length

Name of unit	Equation and symbol
A *decimeter* is one tenth of a meter	$1\,\text{dm} = 10^{-1}\,\text{m}$
A *centimeter* is one hundredth of a meter	$1\,\text{cm} = 10^{-2}\,\text{m}$
A *millimeter* is one thousandth of a meter	$1\,\text{mm} = 10^{-3}\,\text{m}$
A *kilometer* is one thousand meters	$1\,\text{km} = 10^{3}\,\text{m}$

Each of these new units is called a *decimal unit*. We obtain these units by multiplying the basic unit—the meter—by 10^n, where n is a positive or negative integer. This key feature applies to many of other measurement scales of the metric system, **but not to all of them.**

In the U.S., different scales for measuring lengths are also used. The most frequently used units are called the *inch* (in), the *foot* (ft), the *yard* (yd), and the *mile*. 'Mile' is not abbreviated. These units are sometimes called *common English units*. The correspondence between metric units and common units given below is not exact. Note that we use $\approx$ to mean *is approximately equal to*, or for short *is about*.

Name	Equation and symbol
An *inch* is about 2.54 times a centimeter	$1\,\text{in} \approx 2.54\,\text{cm}$
A *foot* is about 0.3048 times a meter	$1\,\text{ft} \approx 0.3048\,\text{m}$
A *yard* is about 0.9144 times a meter	$1\,\text{yd} \approx 0.9144\,\text{m}$
A *mile* is about 1.609344 times a kilometer	$1\,\text{mile} \approx 1.609344\,\text{km}$

Using only common English units:

1 foot is equal to 12 inches	we have[a] $30.48\text{ cm} = 12 \times 2.54\text{ cm}$
1 yard is equal to 3 feet	we have[b] $0.9144\text{ m} = 3 \times 0.3048\text{ m}$
1 mile is equal to 1760 yards	we have[b] $1609.344\text{ m} = 1760 \times 0.9144\text{ m}$

[a] expressing both lengths in centimeters.
[b] expressing both lengths in meters.

The measurement of mass

Name of unit	Equation and symbol
A *gram* is one thousandth of a kilogram	$1\,\text{g} = 10^{-3}\,\text{kg}$.
A *dekagram* is ten grams	$1\,\text{dag} = 10\,\text{g}$.
An *hectogram* is one hundred grams	$1\,\text{hg} = 100\,\text{g}$.
A *decigram* is one tenth of a gram	$1\,\text{dg} = 10^{-1}\,\text{g}$.
A *centigram* is one hundredth of a gram	$1\,\text{cg} = 10^{-2}\,\text{g}$.
A *milligram* is one thousandth of a gram	$1\,\text{mg} = 10^{-3}\,\text{gr}$.
So, a *milligram* is one millionth of a kilogram	$1\,\text{mg} = 10^{-6}\,\text{kg}$.
A *metric ton*[a] is one thousand kilograms.	$1\,\text{t} = 10^{3}\,\text{kg}$.

[a] Or *metric tonne* in the U.K.

Warning. In the U.S., the same name '**ton**' is used in the metric system and in the common English units system, with two different meanings. In the US, a metric ton is equal to 1000 kg, while a '**ton**' is equal to 2000 pounds, which is equal to 907.184 kg. There is no accepted abbreviation for common English 'ton,' while t is used as an abbreviation for metric ton.

Some *common English units* for the measurement of mass.

We recall that $\approx$ means *is approximately equal to*.

Name	Equation and symbol
An *ounce* is about 28.350 grams	$1\,\text{oz} \approx 28.350\,\text{g}$.
A *pound* is about 0.454 kilogram	$1\,\text{lb} \approx 0.454\,\text{kg}$.
A *ton* is about 907.185 kg	$1\,\text{ton} \approx 907.185\,\text{kg}$.
Using only common English units:	
1 ton is exactly equal to 2000 pounds	$1\,\text{ton} = 2000\,\text{lb}$.
1 pound is exactly equal to 16 ounces	$1\,\text{lb} = 16\,\text{oz}$.

APPENDIX

The measurement of capacity

Name of unit	Equation and symbol
A *milliliter* is one thousandth of a liter	$1\,\text{mL} = 10^{-3}\,\text{L}$.
A *centiliter* is one hundredth of a liter	$1\,\text{cL} = 10^{-2}\,\text{L}$.
A *deciliter* is one tenth of a liter	$1\,\text{dL} = 10^{-1}\,\text{L}$.
A *dekaliter* is ten liters	$1\,\text{daL} = 10\,\text{L}$.
A *hectoliter* is one hundred liters	$1\,\text{hL} = 10^{2}\,\text{L}$.
A *kiloliter* is one thousand liters	$1\,\text{kL} = 10^{3}\,\text{L}$.

A few common English units are also used for capacity, which are not part of the metric system. Some of them are the *fluid ounce*, the *cup*, the *pint*, the *quart*, and the *gallon*.

Some *common English units* for the measurement or capacity.

Name	Equation and symbol
One *fluid ounce* is about 30 milliliters	$1\,\text{fl oz} \approx 30\,\text{mL}$.
A *cup* is about 2.37 deciliters	$1\,\text{c} \approx 2.37\,\text{dL}$.
A *pint* is about 473 milliliters	$1\,\text{pt} \approx 473\,\text{mL}$.
A *quart* is about .946 liters	$1\,\text{qt} \approx .946\,\text{L}$.
A *gallon* is about 3.78 liters	$1\,\text{gal} \approx 3.78\,\text{L}$.
Using only common English units:	
1 cup is exactly equal to 8 fluid ounces	$1\,\text{c} = 8\,\text{fl oz}$
1 pint is exactly equal to 2 cups	$1\,\text{pt} = 2\,\text{c}$
1 quart is exactly equal to 2 pints	$1\,\text{pt} = 4\,\text{c}$
1 gallon is exactly equal to 4 quarts	$1\,\text{pt} = 4\,\text{qt}$

The measurement of time

Name of unit	Equation and symbol
A *minute* is $\frac{1}{60}$ of an hour	$1\,\text{min} = \frac{1}{60}\,\text{h}$.
A *second* is $\frac{1}{60}$ of a minute	$1\,\text{s} = \frac{1}{60}\,\text{min}$.
A *millisecond* is one thousandth of a second	$1\,\text{ms} = 10^{-3}\,\text{s}$.
A *nanosecond* is one billionth of a second[a]	$1\,\text{ns} = 10^{-9}\,\text{s}$.
A *day* is 24 hours	$1\,\text{d} = 24\,\text{h}$.
A *week* is 7 days	$1\,\text{wk} = 7\,\text{d}$.

A *year* is 12 months.

A *decade* is 10 years.

A *century* is 100 years.

A *millenium* is 1000 years.

---

[a] It takes about 3 nanoseconds for a beam of light to travel 1 meter.

www.ingramcontent.com/pod-product-compliance
Lightning Source LLC
Chambersburg PA
CBHW051147220526
45473CB00003B/691